Muneo Kaigo
Editor

Social Media and Civil Society in Japan

palgrave
macmillan

Editor
Muneo Kaigo
University of Tsukuba
Tsukuba, Ibaraki, Japan

ISBN 978-981-10-5094-7 ISBN 978-981-10-5095-4 (eBook)
DOI 10.1007/978-981-10-5095-4

Library of Congress Control Number: 2017946017

Cover illustration: jeremy sutton-hibbert / Alamy Stock Photo

Printed on acid-free paper

This Palgrave Macmillan imprint is published by Springer Nature
The registered company is Springer Nature Singapore Pte Ltd.
The registered company address is: 152 Beach Road, #21-01/04 Gateway East, Singapore 189721, Singapore

Foreword

Six years have passed since the Great East Japan Earthquake. The large-scale destruction and devastation from this tragedy has still left many Japanese in shock, as many survivors of this terrible disaster continue to endure hardships and trauma. With the authors of this book, I would like to also express my deepest sympathies to all the victims, and my hopes that all the survivors will soon somehow return to a life that they can call normal again.

Many Japanese had been preparing for a large earthquake for many years, as we all knew it was not a question of if an earthquake will happen, but rather when it will strike. The awful earthquakes in Chile and New Zealand were an eerie message of what was to come. What was different was how the Great East Japan Earthquake was a multidimensional disaster for which Japan was not prepared. Unfortunately, we all learned how all our preparation was not sufficient to counter the size and scale of this earthquake, tsunami, and subsequent nuclear power plant explosion. The contingency plans that were devised by the Japanese ministries and local governments for responding to a large earthquake allowed for many of the Japanese bureaucracies to take the initiative in systematically managing the disaster response after March 11, 2011. Unfortunately, the scale of the earthquake and subsequent tsunami exceeded all of the preplanned responses and this combination led to the Fukushima Daiichi nuclear power plant accident. In the case of the Fukushima Daiichi meltdown, the Japanese cabinet of the Democratic Party of Japan, which was in power at that time, took the initiative in managing the accident without any definite plan and, as a result, was ineffective. This inadequate response will plague Japan for decades to come.

Since 2012, Japan has had a Reconstruction Agency that is in charge of the efforts to rebuild the northeastern region of Japan. The agency also monitors the progress of the decontamination by the Fukushima Daiichi nuclear power plant accident and ensures that no more radioactive contamination occurs in the surrounding areas. Most of the infrastructure has now been restored and the reconstruction of housing is also underway. Although it may take time, Japan is slowly taking steps in the recovery process.

After reading Dr Kaigo's second chapter, I recalled the events that followed March 11, 2011 and that still haunt many in Japan. The widespread destruction throughout northeastern Japan dwarfed what happened in Ibaraki, but the Tsukuba campus of our university also experienced extensive damage on March 11, 2011. At the time, my administrative role as vice-president of the University of Tsukuba required me to help keep the university on course, to allow it to continue operating and to strive so that we could return to a normal state in an extraordinary situation.

Over the course of these past six years, Japan has changed in many ways. Some people have compared the effects of the March 11, 2011 disaster to the havoc of World War II. Although many scholars have already highlighted many of the negative after-effects of this disaster, Dr Kaigo, along with his team for this book, Dr Okura and Dr Tkach-Kawasaki, have been able to highlight some positive changes that have happened in Japan's civil society.

As a scholar, I have also illustrated the importance of civil society throughout my career. In 1997, I initiated the Japan Interest Group Study (JIGS), which is an empirical international comparative study of interest groups and civil society. To date, we have covered 15 nations (Japan, South Korea, the United States of America, Germany, China, Russia, Turkey, the Philippines, Brazil, Bangladesh, Poland, Estonia, Uzbekistan, India, and Thailand). To supplement the findings of this book, I would like to add that I have discovered through my research that the neighborhood associations that we call *Jichikai*, or *Chonaikai* in Japanese, are in fact also rooted in other Asian nations. At the local level, I have found evidence of how these indigenous neighborhood associations have access to the political elites in various nations and are an important element of society. These neighborhood associations have been found to be politically influential in Asian nations, and my research has emphasized how this is a noteworthy difference between Asian nations and others. I was delighted to find that the last two chapters of Dr Kaigo's book makes links to our previous findings about civil society in relation to the dynamics of social media usage.

The neighborhood associations and civil society organizations all made effective use of social media during the Great East Japan Earthquake. As suggested in the conclusion of this book, Japanese civil society has been going through a sort of reawakening with social media. The reconstruction efforts of the northeastern region of Japan have led many Japanese people to rediscover the importance of volunteering. Many people participated in the clean-up of the debris left by the tsunami. Social media shared these experiences of volunteering and led to even more people joining the effort. Japanese college students, which have also had a low presence in civil society for an extensive period of time, have also re-emerged. Twitter and other social media have allowed them to coordinate their efforts with civil society. The project that Dr Kaigo is overseeing is one of those connections between pre-existing civil society and the younger generation of Japanese.

The first two chapters are an enlightening historical journey into the past two decades, highlighting some important aspects of how Japanese society had feared the Internet. And as we all now know, following the Great East Japan Earthquake, many Japanese accepted the Internet as another effective path for acquiring important information. I am certain that the first two chapters will function as a good introduction for anyone who is not too familiar with Japanese media and society to acquire useful insight. The four following chapters focus more on the research and results from consulting and analyzing data from the Tsukuba Civic Activities Cyber-Square, which is a joint project between Tsukuba City, the University of Tsukuba and Intel Corporation. These chapters are useful for understanding methods and strategies for creating synergy through social media with local governments and civil society. They also employ many fresh and original approaches to the study of social media and civil society, and ought to be useful for undergraduate and graduate students that wish to pursue a degree in related fields.

Although I am fully aware that Dr Kaigo is media scholar, I am delighted to welcome this book as a fine addition to the scholarly endeavors of studying the intricate dynamics involved in civil society. I know for a fact that Dr Kaigo has volunteered much time and invested a lot of effort during the past five years as he acts as chair and chief consultant for this project and I am delighted to see the fruits of his research presented here. I hope that this book will inspire interest into research of civil society and also function as a permanent reminder of the disaster that has changed Japan so greatly.

Yutaka Tsujinaka
Professor, University of Tsukuba
Tsukuba, Ibaraki, Japan
Director, Institute for Comparative
Research in Human and Social Sciences
University of Tsukuba
Tsukuba, Japan
March 11, 2017

Preface

The main objective of this book is to summarize the findings of a large-scale experiment in Japan designed to connect civil society organizations that were already active to become increasingly interconnected online. To be quite frank, this is more or less the exact opposite of most approaches to research in this field and the authors believe it to be fairly original, as it is attempting to do the reverse of the more usual approach. For instance, when you search for research on social media and civil society, most research is focused on how to have people who are already online, to engage in real life.

This second half of this book covers the analysis results of a five-year research project designed for connecting civil society organizations, citizens and local governments through online communication via social media for the promotion of higher levels of citizen engagement in Japan. The book not only relies on traditional hypothesis testing and social surveys, but also employs more familiar data in the realm of marketing and also takes on large-scale (big) data analysis. Therefore, the second half of the book focuses on Japanese civil society and the social media scene.

The project was fielded at the Tsukuba Science City in Ibaraki Prefecture, Japan located 60 kilometers north of Tokyo and is currently being managed through the municipal government of Tsukuba City. Among the majority of the Japanese local government Facebook pages that focus mainly on tourism or promotion of local industry, the Tsukuba Civic Activities Cyber Square is the most successful and largest online community in Japan that focuses solely on civic activities with a respectable level of online engagement. The book hopes to provide insight into the development of online communities for connecting citizens already participating in civil society activities such as volunteering. This book is also valuable as it also attempts to explain how to create, manage, nurture and grow online communities that are focused on non-viral civil society themes such as volunteering. Through a careful examination of the multi-year project, it captures the challenges and opportunities in designing such communities that are aimed at connecting offline groups via social media such as Facebook.

The main foundation of the project was conceived during and after the Great East Japan Earthquake which devastated large portions of the northeastern area of Japan. The project latently aims to prepare for a more resilient communication network among citizens in case of another large-scale disaster. The book targets professionals interested in the subject matter, but also undergraduate and graduate-level students.

The first chapter offers readers a good introduction to the unique Internet environment of Japan and also offers better explanation of the cultural norms that surround Japanese social media. This chapter focuses on the elements that contributed to the hesitancy and the sluggish adoption of ICT, the Internet and social media in Japan. I explain the dynamics of low self-efficacy (or *Nigate-Ishiki*) in Japan, media access and how other various elements have contributed to shying away from using the Internet.

In the second chapter, I make a general overview of social media usage in Japan and describe the background of the project through a recap of the role of social media during mid-level disruptions by natural disasters by examining the case of the city of Tsukuba in Ibaraki prefecture during the Great East Japan Earthquake of March 2011. The chapter also reviews the role of the Fukushima Daiichi nuclear power plant accident and how that accident led to more Japanese to adopt social media.

In Chap. 3, Sae Okura and I conduct a text mining of all available 425 Facebook pages run by Japanese local governments. This chapter examines how these SNS pages are being used by local governments to create a more collaborative relationship with the private sector. The chapter also investigates what policy areas tend to get more fans or followers and engagement on Facebook. The result here presents how the number of fans or followers and level of engagement of Facebook pages run by Japanese local governments are determined by policy areas.

In Chap. 4, Leslie Tkach-Kawasaki and I introduce the Tsukuba case study and social media usage. It discusses the possibilities and problems of complementary communication channels such as social networking services for promoting civil society activities and linking civil society organizations. The chapter focuses on the first phase of the Tsukuba Civic Activities Cyber-Square (Tsukuba Shimin Katsudō no Hiroba) on Facebook Experiment from early in 2012 and how it functioned during and after the May 6, 2012 Tsukuba city tornado disaster for subsequent relief and support activities.

In Chap. 5, Sae Okura and I conduct a qualitative analysis of the interviews with the community managers of the Tsukuba Cyber-Activities Square. The analysis results explain how the daily procedures of the local government interrupt the development of the Tsukuba Civic Activities Cyber-Square Facebook page. The chapter explores how government-initiated computer-mediated communication (online activities) and human communication (offline activities) increases the level of public engagement and also, on the other hand, how some activities have a negative impact on the level of public engagement on the Tsukuba Civic Activities Cyber-Square Facebook page.

In Chap. 6, Sae Okura and I examine the advocacy networks of civil society organizations (CSOs). Although the importance of the activities in advocacy by CSOs in policy and decision-making procedures has been greatly emphasized in the literature of political science and social policy, we have relatively little understanding of the relevance and impact of the leading actors that structure the diverse networks and discourses through social media. This chapter analyses CSOs at the local government level involved in advocacy activities through the use of social media such as Facebook and Twitter. Based on the "Tsukuba Civic Activities Cyber-Square," the chapter explores how social networking services such as Facebook can provide civil society organizations with: (1) more political opportunities to advocate; (2) more chances to connect with the local government; and (3) create opportunities to exert greater presence, in spite of their limited financial and political resources.

In Chap. 7, the conclusion, Sae Okura and I review the findings of the previous six chapters and examine how the Japanese adoption of social media increased during and after the Great East Japan Earthquake and the Fukushima Daiichi nuclear power plant accident as both of these catastrophes acted as catalysts for this process. We also created hypotheses for further research. One potentially important hypothesis is a possible number determining the lower threshold for maintaining a virtual community to be self-sustainable. We also list the accomplishments of the Tsukuba Civic Activities Cyber-Square and explain how the introduction of social media into the local government of Tsukuba City has changed the attitudes of municipal workers, civil society and has led to the regeneration of new civic activities.

Muneo Kaigo

Acknowledgements

This book is a based on the results of a five-year project that was designed to connect civil society organizations, citizens and local governments through online communication via social media for promoting higher levels of citizen engagement in Japan. As there were no proper manuals on how to make this type of project successful, to be honest, it was quite challenging. In the process, the project became truly multidisciplinary, as we used theories and methodology from various fields. After the third year, we were beginning to see some very positive results, and by the end of the fifth year, the project has been quite fruitful for everyone involved. I am grateful for all the participants in this endeavor, and the people who are in the Tsukuba Civic Activities Cyber-Square community.

Partially based on numerous previous conference papers (ICA, IAMCR, CeDEM, ITS, JSICR and AMIC) and/or content shared with the academic community as working papers, university bulletins ("Social Media Usage During Disasters and Social Capital: Twitter and the Great East Japan Earthquake" *Keio Communication Review 34*, pp. 19–35. March 2012.) or papers shared through the Creative Commons license ("An Analysis of Japanese Local Government Facebook Profiles" *CeDEM Asia 2016 Proceedings of the International Conference for E-Democracy and Open Government,* pp. 67–80, 2016; "Who Leads Advocacy through Social Media in Japan? Evidence from the 'Tsukuba Civic Activities Cyber-Square' Facebook Page" *Information 7*(4), 2016; "Social Media for Enhancing Civil Society and Disaster Relief: Usage by Local Municipalities in Japan" *eJournal of eDemocracy and Open Government 7*(1), pp.1–22, 2015; "Internet Aggregators Constructing the Political

Right Wing in Japan" *eJournal of eDemocracy and Open Government* 5(1), pp. 59–79, 2013), so we have newly written, rewritten or revised all portions of this book.

Our work has been supported by JSPS KAKENHI Grant Number 25330394 and 18500191, and the Institute for Comparative research in Human and Social Sciences of the University of Tsukuba.

Sae Okura was a key force over these past five years in helping me take on this project. She began to help me with the project while she was still working on her dissertation under Professor Tsujinaka as a graduate student. Her contribution with the latter half of this book was vital. I would like to thank all the people of the Tsukuba municipal government, as well as Intel Co., for assisting and cooperating with our research. I would also like to thank Dr Yutaka Tsujinaka, the director of the Institute for Comparative Research in Human and Social Sciences of the University of Tsukuba, for writing the foreword to this book. Leslie Tkach-Kawasaki was kind enough to help me with Chap. 4. Yohei Kobashi and Yoshiaki Kubo were important keys in assisting and developing our research. I would also like to thank Teruyoshi Sasaki and Jun Oguro for their insights and inspiration during my research.

From the Graduate School of Humanities and Social Sciences Doctoral Program of International and Advanced Japanese Studies, I would like to thank Dragana Lazic, Natalia Novikova and Lisander Martinez for helping us with the indexing of the book. From the graduates of the MA Program, I would like to thank Haruki Tajima and Wonyoung Jang for their assistance with collecting data for the project. I would like to thank Bob Kaplan for teaching me the art of perseverance in research and writing.

Finally, I would like to express my sincere appreciation to Sara Crowley Vigneau for making this book a reality and my gratitude to Connie Li, and thanks to M. Vipin Kumar and his team and all the people at Palgrave Macmillan and Springer Nature involved in editing and overseeing this book through the publication process.

March 2017 Muneo Kaigo

Contents

List of Figures

List of Tables

CHAPTER 1

The Japanese Internet Environment

Muneo Kaigo

Among a good portion of the Japanese population, the Internet is perceived to be a convenient, but scary world, a place of deception and fraud. Many Japanese are cautious when using the Internet because it is seen as a place where malicious people are prowling around, filled with the utmost horrifying information or indecent images, and should either be used with great caution or simply avoided as much as possible. Others are cautious or paranoid about using the Internet because they believe that devices can easily become infected by a computer virus or malware. Although this image of the Internet is a somewhat extreme one, all of these concerns are well grounded in Japan, and this leads to further discomfort when it is necessary for Japanese people to get online.

The Internet is a global network that reaches out to all parts of the world, a terrifying prospect if one is xenophobic, and quite intimidating for many Japanese who feel discomfort when using English or indeed languages other than Japanese. In comparison to the relatively safe, everyday life in Japan of social law and order kept throughout most of Japan with its flow of sanitized mass-mediated information, cyberspace provides a break into an area of freedom for everyone and everything that does not

M. Kaigo (✉)
Faculty of Humanities and Social Sciences, University of Tsukuba,
Tsukuba, Ibaraki, Japan

M. Kaigo (ed.), *Social Media and Civil Society in Japan*,
DOI 10.1007/978-981-10-5095-4_1

belong in the Japanese social environment. For some, the Internet provides a fresh space of liberty that was hitherto unattainable in the traditional Japanese social context. However, for many Japanese, the Internet is viewed as a dangerous area that turns everyone into vulnerable prey for the vicious cyber criminals and trolls. The prevalence of security incidents that are portrayed in the Japanese mainstream mass media adds fuel to these concerns.

One may attribute this perception among the Japanese to be the reason why the diffusion of Internet usage in Japan among the general population during these past few decades had been relatively slow when compared to many other developed nations. However, other underlying elements have contributed to the sluggish diffusion of the Internet in Japan and have led to a so-called digital divide in Japan as well.

Nigate-Ishiki: Avoiding Computers and the Internet in Japan

Research on the digital divide has traditionally been more focused on addressing the socio-economic factors that encompass differences in attaining information among races or classes in society (Hoffman et al. 2000; Van Dijk and Hacker 2003). However in the case of Japan, racial diversity is not so much of an issue. Economic capabilities for acquiring new information and communications technologies (ICT) is also not a grave issue, although historically the costs of being online, and indeed of telecommunications in general in Japan, has been relatively more expensive than in other developed nations. This cost factor may have slowed down quicker and wider Internet adoption in the home until the early 2000s in Japan. What can be considered a more significant issue is the combination of the culturally distinct affective and cognitive factors that have influenced the adoption of new ICT among Japanese in the past. This combination may be one important element that has also influenced the perception of the very nature of the Internet itself among Japanese over the past few decades.

Much of the traditional interpretation of how humans recognize and process their environment has been based on many fundamental concepts of cognitive psychology. These concepts have been derived from advances in computer science and the results have been applied in examining how humans process information cognitively (Simon 1979; Anderson 1980; Norman 1981). Therefore, the understanding of these human processes has been concerned with the understanding of computer science and also research into artificial intelligence. However, this aspect might have actually been unintentionally inhibiting advancing research into how humans deal with information and communication technology from different perspectives. A broader interpretation of the communication processes involved in the use of information and communication technology can shed light on the sluggish adoption of the Internet in Japan and also illustrate some of the more relevant human communication and cultural factors that have been relatively unexplored.

Excluding the Barker and Edwards (1980) intrapersonal communication model, few models have dealt with the intrapersonal feedback within the encoding process of communication. Many of the communication models that exist now, have perhaps unintentionally omitted details of the intrapersonal feedback within the encoding process of language. Before the advances in the technology of voice recognition or telepresence which has greatly changed currently what is considered ICT, Rogers (1986) had stated how communication through communication technology was mainly asynchronous. As this idea still holds true in many features of using ICT, the loop of intrapersonal feedback is relevant and should be reviewed more carefully. In the process of asynchronous communication of ICT, one is constantly using a loop of intrapersonal feedback to improve the fidelity of communication. For example, when one writes a letter to a friend, intrapersonal communication is being conducted within the writer and in the same way, a loop of intrapersonal feedback occurs when conducting asynchronous communication with ICT. In order to send the intended information with the least amount of distortion to the receiver, one will continue this internal encoding feedback process while building the content of the message.

In consideration of this notion in the Japanese context, one realizes that it is important to focus on the encoding–feedback process of ICT. In fact, among most areas of the globe using languages that are non-alphabetic, the encoding–feedback loop is a salient feature of communication through information and communication technology. In these areas, input methods utilize two-bit (two-byte) encoding to process characters for each distinct language (Nishigaki and Lewis 2001). Two-bit encoding is the process in which the characters displayed are not using simple one-bit ASCII characters available on the keyboards but employ two-bit characters to formulate one character, like the Japanese front-end character processing. In this interface environment, the feedback loop should be recognized as an important aspect of communication through ICT as this clearly is different from typing one-bit ASCII characters, as in English. The awareness of this feedback loop in ICT use validates viewing the use of ICT as an independent component of the communication process and deepens our understanding of modern communications as a whole.

One concept that can be suggested to directly influence this encoding–feedback loop during initial ICT usage is self-efficacy. The concept of self-efficacy (Bandura 1995, 1997) examines how one formulates the belief in maintaining confidence in a skill, and the managing of one's capabilities to execute a certain action. A long line of computer anxiety studies has continued evolving, and it has merged with the research of self-efficacy in examining new ICT use and apprehension. Along with the indications made by Bandura (1997), Eastin and LaRose (2000) in the past had conducted a study that examined Internet self-efficacy and they explained many of the initial barriers involved in Internet use as self-efficacy deficits, or low self-efficacy. However, research into the different factors involved in users' ambivalence towards computers is nothing new. Prior to this, many studies focused more on the relationship between information technology and anxiety, especially concentrating on computer anxiety. Along with anxiety, many other variables have been examined in the past, for example, Paxton and Turner (1984) examined human factors such as inexperience, needs, and anxiety involved when using a computer. Gilroy and Desai (1986) found that gender differences might affect anxiety and, in addition, that different training methods can be used to reduce anxiety. Parasuraman and Igbaria (1990) and Igbaria and Chakrabarti (1990) further analyzed the relationship of computer attitudes, anxiety and gender. They discovered that among their samples, gender and anxiety were the strongest predictors of attitudes towards computers along with other determinants. Dyck and Smither (1992) conducted research on the relationship of computer anxiety and experience, gender, education and age. Their sample of younger and older adults confirmed a negative relationship between computer anxiety and computer experience, and also found that older adults with less computer experience were less apprehensive towards computers. Szajna and Mackay (1995) found that computing aptitude and achievement are related to the user's learning performance, however anxiety and experience are not. Day and Makirinne-Crofts (1997) examined the interface features that contribute to computer anxiety. They investigated constructs including cultural and individual differences, interface quality, self-efficacy, ease of use, user attitudes and intended usage behavior. Chua et al. (1999) conducted a meta-analysis of computer anxiety and have found that computer anxiety leads to computer avoidance, but computer experience also leads to a decrease in anxiety. They added that computer anxiety is a kind of "state anxiety" which can be altered, and that such anxiety is perceived differently by people from different nations. The relationships between computer anxiety and gender were found to be inconsistent, but age was related to computer experience and is inversely related to computer anxiety. By investigating the results of these past studies on computer anxiety and self-efficacy, a variety of variables have been found to exist but many of the studies are focused mainly on computer anxiety (Bush 1995; Whitley 1997). However, in consideration of the findings in Eastin and LaRose (2000), the concept of self-efficacy with ICT and the past results in computer anxiety research can be merged to explain some uncovered mechanisms that can be found in what was delaying adoption and initial use of computers and ICT in Japan. Integration of such research for recounting the Japanese situation is viable, although these past concepts need to be reformulated because fear or technophobia, technology apprehension are all too strong to explain the dynamics involved in the Japanese context.

As mentioned earlier, the Japanese encoding–feedback process has been an important aspect among the various processes of communication when using ICT in the Japanese environment and the process of ICT usage is directly influenced in this process. Computer-mediated communication traditionally adopts a human–machine interface in the message production process, so it is a veritable encoding–feedback process by the communication source. During this process, the Japanese construct of "*nigate-ishiki*" can comprehensively explain the evident self-efficacy deficiency (Eastin and LaRose 2000) and aspects that lead to avoidance of computers and the Internet in Japan in the past.

The construct *nigate-ishiki* that is a sentiment or consciousness of naturally being bad at dealing with something or someone, which leads to uneasiness, shyness or avoidance. The construct can be useful in explaining the mechanism affecting the source–encoding feedback through computers in Japan. More importantly, this construct can also can explain the sluggish adoption of social media was affected as well in Japan as we see later in this book. From the meaning of the construct, *nigate-ishiki* has common features of Bandura's (1997) notion of low self-efficacy; however, one can observe how this trait is developed from the common method of how people are evaluated by others in Japan. As a general rule that can be observed most prominently in the career and life-determining entrance examination systems in Japanese education, Japan as a society is more "demerit focused" (people are scrutinized for mistakes or what they cannot do) than "merit focused" (people commended for what they can do). The conditioning among the youth of this feature has common elements with the facet of self-efficacy and academic anxiety (Bandura 1997, p. 77, p. 235). However, this does not imply that *nigate-ishiki* is only common among children. In fact, when age is taken into account, *nigate-ishiki* is more prevalent among adults than it is among children, due to the traditional Japanese cultural tendency of the respected to "save face" when that one is in an awkward situation (Ting-Toomey 1988). The mentioning by a senior aged person having *nigate-ishiki* serves as a non-imposing excuse to avoid various things, people or situations, and the excuse of *nigate-ishiki* also saves one's face. *Nigate-ishiki* can serve as an excuse for an individual in an imposing situation and in some cases can be forgiven from disgrace, rather than being "crucified" as an "incompetent" individual lacking the competency to be able to do something. In this sense, another facet of *nigate-ishiki* is common with some the notions of social anxiety, self-efficacy and control (Bandura 1997; p. 323). *Nigate-ishiki* has the nuance of not having confidence in something and wanting avoidance but it is not totally compatible with the features of anxiety or apprehension, that include stronger emotions such as fear. This behavior of avoidance may apply to things but may also apply to avoiding specific individuals or people and due to not being based on a dual process theory of anxiety-avoidance, it is in accordance with Bandura's (1997; p. 324) assessment of avoidant behavior.

The construct of *nigate-ishiki* of ICT usage can be explained through the ambivalent relation between confidence (one facet of self-efficacy) and the level of anxiety involved. Although some anxiety exists in use of ICT, it is more of an ambivalent state that balances out with lack of confidence. This ambivalent relation between the conflicting emotions of self-declared competence and anxiety towards ICT explain *nigate-ishiki* towards ICT. *Nigate-ishiki* is a potential key in explaining the delay of Japanese ICT adoption in the 2000s and may also be one contributing element to explain the persistent reluctance of many Japanese to adopt newer means of communication technology.

As mentioned earlier, the slowness in adopting an efficient man–machine interface with ICT such as computers has been due to the language-character incompatibilities (49 basic characters in the Japanese language) and as a result, many Japanese businesses, to this day, still continue to have a heavy dependence on photocopying and facsimile communication technology, especially those that have an older clientele such as the real estate businesses. However, Japanese perceive that a cultural inclination towards the English-language culture exists when one considers the human–computer interface (such as the ASCII keyboard and other computer system-related jargon) of computers. For example, a large portion of the currently used Japanese computer terms are direct phonetic imports from English, and are therefore assigned sounds and characters (*katakana*—the Japanese writing system for words that are phonetically imported from outside of Japan) most similar to the original English term. In result, Japanese computer terms inherently lack the context embedded in the original English operating system terms, such as: file, cut or paste. Through this, one can assume that it also cultivates an intercultural aspect of *nigate-ishiki*, and such intercultural tension might be evident especially when the person adopting the technology is not an adept user of the English language. This tension caused by the interfacing language can be explained through the tension caused by this "intercultural" encounter.

There have been previous examinations of Japanese cultural characteristics in ICT adoption. In a previous survey with my colleague, we empirically found an inverse relation between confidence and anxiety, and a high correlation between age and confidence, and age and anxiety (Kaigo and Sasaki 2001). This survey result demonstrated an ambivalent relation between confidence and anxiety when putting the perspective of age into account and explains one characteristic of explaining the reasons for reluctance of Japanese ICT adoption. The way that both confidence and anxiety increase with age is difficult to explain solely using concepts of self-efficacy or anxiety, but with *nigate-ishiki*, one can explain how the older respondents in our survey, who were mainly elementary and junior high school teachers, experienced a certain amount of ambivalence with regard to ICT. To further explain this dynamic, in the Japanese educational situation, the teacher must have control of all aspects of the classroom and must also demonstrate a high level of confidence in a subject. This includes ICT. However, at the same time, if the teacher is not highly adept, he or she is likely to experience a high degree of anxiety. A senior aged person will not like to admit being bad with computers, but may confess to experiencing some *nigate-ishiki* towards ICT. When Weil and Rosen (1995) conducted a survey among Japanese students concerning computers and anxiety, they indicated that very high levels of anxiety were found in the results in comparison with other nations and they labeled their findings concerning Japan an "enigma." The concept surrounding ***nigate-ishiki***—instead of "fear"—might have been the "key" to their "enigma" of Japanese with low self-efficacy towards computers.

Avoidance of communication through ICT in Japan in the past has been common in three areas: (1) problems in prior experience with ICT leading to apprehension; (2) the preference of written communication; and (3) dexterity problems. To explain what the problems in prior experience were; along with ***nigate-ishiki***, Japanese also expressed computer anxiety traits, such as the apprehension of breaking or damaging computers because they would freeze all the time. In the past, some computers had an overload of CPU and caused erratic performance due to the burden of having to display the Japanese language. A general preference of written communication might be a distinctly cultural aspect of older Japanese that has led to the convenient avoidance of ICT as they would express that a word-processed Japanese document printed out on paper seemed "inhuman" and "cold" whereas written Japanese is more "warm" and aesthetically preferable, especially when transmitting emotional messages. Many companies in Japan prefer the curriculum vitae to be handwritten by the college students who strive each year to get recruited by the company of their preference. The dexterity problems experienced by senior citizens in Japan have mainly discouraged them from trying to use the computer interface, but their problems are also with other devices as they have mentioned how mobile phone buttons and the screen displays on mobile phones are too small for them to use and see, and the graphic user interface (GUI) computer icons are difficult to recognize.

The advances in new ICT with new communication devices and network infrastructure have made communication through new ICT unavoidable in many everyday situations. As in the way computers were introduced into homes, social media is also becoming an important and vital information exchange method in many nations. Nevertheless, smartphones have been leading a large portion of the Japanese population into Internet usage. Even with low computer skills and literacy, smartphone users are able to manage basic communication skills involved in using the Internet such as email and web browsing.

The past focus of research in the past on the personal computer or mobile phones is inevitable when considering their roles in the diffusion of ICT. As we have seen in these three areas in Japan that have led to the avoidance of ICT, the larger problem remains in the older age group of Japanese that continue to experience ***nigate-ishiki*** towards smartphones to this day. The senior generation of Japanese have trouble when attempting to use smartphones as they find it difficult to manipulate the devices, even though the interface in language input has improved greatly. Haptic control of the devices is difficult for those with less dexterity and the App icons are difficult to distinguish when age contributes to many vision-related problems. If the persistent *nigate-ishiki* among the older generation of Japanese cannot be resolved, the residual sluggishness of adoption of ICT and social media in Japan may be "sowed" here, where many will be labeled non-literate in ICT and therefore not fully functional in the new Japanese digital communication environment which involves the widespread use of social media.

Media Access: Digital Skills and Smartphones in Japan

In previous observations of the elements related to the gap between those who have access to network information and not, Newhagen and Bucy (2004) indicated how it is important to have physical access to networks, but the ability to freely access content and information on the Internet is even more important. Van Dijk's (2004) ideas of how to assess an individual's ability and skills for accessing networks is unique and is aptly named *digital skills.*

According to Van Dijk, digital skills are those abilities that define how one operates hardware and software—instrumental skills, selects, processes and applies information from digital sources—informational skills and to strategically uses them to improve one's position in society—strategic skills. In comparison to the term computer literacy—which includes such abilities as constructing computer programs—digital skills refer to the ability to have access to networks, therefore it is more relevant to the actual skills necessary for those functioning in the growing digital environment.

The importance of using digitalized information is obvious in our societies now, and whether individuals are capable of accessing various information through the Internet becomes increasingly important. Speed and control, self-efficacy can be considered elements that determine how an individual can actively participate in the digital environment. In this environment, can the concept of digital skills apply to Japan and provide a view on how ICT can be smoothly adopted for other nations that use a non-alphabet based language? The Japanese experience from this perspective may prove useful for other nations that do not use a non-alphabet based language and are still in the works to become a networked society.

Van Dijk's concept of digital skills deals with four kinds of new media access: (1) mental access—lack of elementary digital experience caused by mental factors such as computer anxiety, lack of attractiveness, (2) material access—not having possession of computers or connections to a network, (3) skills access—the lack of digital skills caused by interface problems or formal and or informal education and (4) usage access—lack of opportunities for usage. For the Japanese situation, (1) mental access and (3) skills access are larger problems when compared to (2) material access or (4) usage access, which are evidently larger problems for other nations.

Upon considering the Japanese situation, digital skill elements need to be considered in tandem with what are known in Japan as *mind elements* or the basic traits necessary for living in the information society, the proper state of mind that one needs in order to succeed in a networked society (Yotsumoto 2002). The mind elements defined by Yotsumoto have been described as the following traits and abilities: (1) motivation for self-actualization or the drive to use time for improving one's lifestyle, this trait requires one to having a clear picture of an ideal future "self" and actually working towards this; (2) societal involvement and being involved in society, willing to help other individuals, maintaining awareness of what is happening in society by keeping up-to-date with news or documentary programs; (3) accepting the price for important information, and willingness to pay for information is necessary in modern society; (4) being inquisitive of new things, new information and having a basic understanding of important agenda in society; (5) being able to organize and come to conclusions by interpreting large data, and being evaluated by others as an adept teacher of things; (6) being enthusiastic of sending out ideas and opinions, being able to positively accept criticism; (7) being able to communicate with persuasion, always trying to use easy to understand examples, and simple conclusions when communicating; and, finally, (8) being capable of providing useful advice to other people, sharing ways of how to do things and asking for the opinions of others (Yotsumoto 2002).

These mind elements complement the concept of Van Dijk's strategic skills of digital skills when overviewing the Japanese situation. In the case of examining Japan, the interaction of mental access and skill access is related to how language abilities affect digital skills. Through this interaction, the instrumental and informational skills among the digital skills conceptualized by Van Dijk are most affected.

As we have discussed earlier, Japanese is a language that is not alphabet based, but translation software programs can now allow the wider reception of content as such programs allow information in other languages to be understood through one's native language. Unfortunately, because of the difficulties in the procedure of translating Japanese and most other languages, translation software between Japanese and other languages still result in comparatively crude translations in comparison to, i.e., English—French translations, resulting in lower fidelity requiring more reinterpretation by the individual and requires more for actual interpretation of information. Hence, the speed of access to certain information is affected due to the cultural barriers made by the Japanese language. This can potentially influence the level of informational skills of Van Dijk, because language is the basis for the ability in searching, selecting, processing and applying information from digital sources. Obviously, the focus here is not on language itself, but more on *which* language is normally being used for informational skills. As English is one of the most common and inherently dominant languages of the Internet, whether one is capable or has facility in comprehending and processing English determines how widely an individual is able to access different textual information. The content that is being translated into Japanese or any nation's language determines what information is available to the majority of those that can only understand one's language. The importance of content information access with instrumental purposes is an essential aspect in closing gaps in access. We shall later see in Chap. 2 how this affected the reception of important information during the Great East Japan Earthquake and the Fukushima Daiichi nuclear disaster.

Nowadays the digital divide is no longer deemed as an issue in Japan as the rapid diffusion of smartphones in Japan seems to have closed these gaps in access. However, when content variety accessed through smartphones is measured, one can easily discover that the most commonly accessed information is limited and the smartphone is not giving access to the wealth of the Internet. Limited content can be accessed through these mobile-oriented websites and this results in limiting smartphone information access and becoming largely entertainment oriented. Smartphones, which are undeniably rapidly advancing ICT accessibility, have limitations in terms of capability of access when compared with personal computers. In result, this limits the type of access to content available on the Internet because access to the Internet through smartphones tends to have more entertainment purposes with an "immediate reward" that can be gratifying to the user without mental effort. In result, a latently dysfunctional aspect of content access becomes evident where smartphone optimization may be creating another gatekeeper of information access.

From the perspective of media access, one can suggest that the pursuit of instrumental purposes among all information access is what needs to be made a priority. The smartphone is an ideal medium to allow those with *nigate-ishiki* towards other ICT to stay online. This medium has developed into an essential digital tool in the everyday life of Japan. In the same way that television and computers are accompanied with different orientations, smartphones and computers have different orientations, and the ease of access of various information is different for each medium. As entertainment-oriented content with immediate reward is more commonly accessed through smartphones, they have succeeded in providing ICT for many. As full access to the wealth of content and information that is available on the Internet becomes increasingly important, one will realize that it may still take a little more time for smartphones to replace computers as the sole digital devices that allow instrumental purposes of Internet use. Tablets may eventually enable this feature, however smartphone use with instrumental purposes will be difficult unless the characteristic limitations of the medium evolve further. This trend indicates that when content access is conducted through limited hardware, it leads to limited access.

The factor of technology itself is not a problem in the adoption process of ICT because one can observe, e.g., how gaming technology has diffused over the years. In the case of Japan and perhaps other nations, the larger problem is how Internet access with instrumental purposes with "delayed" rewards requiring more mental patience in general tends to be low. The reality-oriented instrumental aspect of new ICT use for active content access might directly affect self-efficacy in new ICT use. Content in text—which is predicted to remain dominant in the Japanese Internet user environment for the time being, requires the individual to be more active within the cognitive domain of the mind and poses a larger problem in perception of the Internet in Japan when compared to other nations.

For Japanese to overcome this, more development of the affective domain of the mind, for example curiosity, becomes necessary to drive the mind into active access to content. In recent years Japanese education has been unsuccessful in fostering the affective side of learning and the policy makers in Japan should remember that one main feature of education is not to teach, but to foster and develop the affective side of the learner's mind as well. Furthermore, a radical change in perception of how to teach and learn foreign languages and cultures urgently awaits implementation in Japan.

Until new general policy measures are implemented, a type of classification based on digital skills, divided by technology and lifestyle, will determine the level of access to information through the Internet. As smartphones have reached high adoption rates in Japan, there will be less of a digital divide or "haves and have nots" concerning ICT, but gradual shades of different classes among users based on digital skills, foreign language comprehension strategies, the orientations of users and means or devices of Internet access. Elements to determine the class of each individual in this sort of digital classification can be readily assembled according to Van Dijk's comprehensive categories of access: mental access, material access, skills access and usage access. The degree of classification will be defined according to the breadth of information that is accessible to each individual. Among the Japanese, the ability of digital skills with foreign, but in particular English language skills, stands out. Language ability can change the level of control of ICT access to information, as those who are uncomfortable with foreign languages will have limited access and those who cannot use any foreign languages will be even more limited in access. Accompanying these elements, access to personal computers or smartphones will determine the variety of information that can be accessed. Material access will determine the breadth of access, where there will be those who will have superb access with the best hardware, top speed bandwidth capable of accessing the best broadband content, and speed for gaining information through information and communication technology, and there will be those who will have very limited access at the bottom.

In Japan, the highest classification possible will be among those individuals with optimum access through various devices with high digital skills and foreign language facility. They will have access to the widest variety of content. The individuals with optimum conditions but with *nigate-ishiki* in digital skills and or foreign language will be on a lower classification. Translation software can be a complementary means to close the gap with these individuals to a certain extent. Even those who are economically capable of acquiring optimum conditions but only having smartphone access will also be on a lower position of this classification because the conditions for optimum access are not obtained.

Furthermore, one can predict that the disadvantage will remain for those further on the lower end, because limited network access in terms of language and hardware may lead those on the lower end of the classification to only access entertainment. Smartphone access to the Internet with single language capability is analogous to standing in front of a magazine rack in a convenience store or newsstand. On the other hand, the content available on the Internet through optimum conditions of a personal computer with multiple language facility is more similar to having access to a vast library of information.

Zeno's paradox is a faulty illustration of how, in a race, Achilles will never be able to overtake a tortoise. In the paradox, Achilles is in a race and runs after a turtle, but can never overtake it. The mathematical interpretation of this paradox is as follows: For by the time (t_1) Achilles reaches the spot where the turtle was initially (at time t_0), the turtle has crawled forward to another spot. By the time Achilles reaches that spot (at t_2), the turtle has crawled forward to another spot.

$$\begin{aligned}
A &= \text{Achilles' speed} \\
T &= \text{turtle's speed } (A > T \text{ presumably}) \\
L &= \text{turtle's initial lead.} \\
t_1 - t_0 &= L/A \\
t_2 - t_1 &= L/A \times T/A \\
t_3 - t_2 &= L/A \times T/A \times T/A \\
&\cdots
\end{aligned}$$

Adding, the total time elapsed by the nth step is

$$\begin{aligned}
t_n - t_0 &= L/A \times \left\{1 + T/A + \ldots + (T/A)^{(n-1)}\right\} \\
&= L/A \times \frac{1-(T/A)^n}{1-T/A}
\end{aligned}$$

Since $0 < T/A < 1$, $(T/A)^n$ converges to 0 when n goes to infinity, and the whole expression converges to $L/(A - T)$. In other words, after time $L/(A - T)$ Achilles catches up with the turtle. Their relative speed is $A - T$, so the time it takes Achilles to overtake the turtle's initial lead L is $L/(A - T)$. The paradox arises because some believe that if you add infinitely many terms, the sum must also be infinite and that is incorrect. One can add infinitely many, but smaller and smaller terms, so as to make the sum finite.

This faulty paradox can metaphorically describe the problems in Japan with regard to Internet connection through limited technology and skills and can also be used to explain the digital divide among nations. Metaphorically speaking, the ownership of mobile phones or smartphones with low or limited digital skills can be compared to Achilles' whereas personal computer usage with high skills can be viewed as the tortoise of the paradox. The national statistics in Japan display that the adoption of mobile phones is complete and, as a result, Internet adoption is also complete in Japan. From one point of view, the smartphone (Achilles) has already triumphed over the personal computer (tortoise). However, when considering access to content, the smartphone user with limited digital skills does not fit the definition of proper media access, in such conditions as can only be provided through personal computers and higher levels of digital skills. In this sense, the smartphone has not overtaken the personal computer and therefore metaphorically, the paradox is substantiated here.

The continuation of further implementation in resolving *nigate-ishiki* of ICT among Japanese individuals should continue to be implemented. At the same time, however, more importantly, self-efficacy in the ability to adopt new ICT should be fostered and focused on, more than actual self-efficacy in current ICT. In other words, the self-efficacy in having the abilities to periodically realign oneself to the newest mode of optimum access through information and communication technology should be focused on.

Newer platforms of communication and new technology will continue to be developed. The currently dominant social media platforms may become obsolete in the next decade. However, as social media is becoming an important way to exchange information among those who are online, one must examine on how this platform can benefit society as we shall see in the later chapters of this book.

Infection: Viral Contagion Through Sharing Files over the Internet

Sharing files over the Internet, such as music or movies, is more often than not an illegal and a disputed activity that is engaged in many areas that have high bandwidth Internet connections. The debate over downloading media files is both a social and public policy issue, and the legal and economic aspects have been disputed from both sides in various nations. In Japan, Winny was for a time the most infamous peer-to-peer (P2P) platform and operates on the Microsoft Windows OS environment. The software was developed by a former University of Tokyo assistant named Isamu Kaneko.

Information about this software was exchanged or distributed through the main/largest Japanese Internet forum called the *Ni-Channel* (2Ch).

The *Ni-Channel* or 2Ch is infamous for its many problems and distasteful posts (Onishi 2004). Indeed, given all its counterproductive aspects, the *Ni-Channel* is a well-known chaotic community in Japan. Many topics that are avoided in daily communication in Japan are among the most popular topics to be found in *Ni-Channel*. Another feature of the *Ni-Channel* is the frequent and vulgar incivility among users, contrary to a culture where face-to-face conflict is unwelcome and is infrequent in formal communication among Japanese. *Ni-Channel* has also become a vortex for cultivating hate speech, and a place where criminal activities have taken place. The anonymity of the users inherently hides information of who is posting something and, as a result, the content of *Ni-Channel* often appears as the worst examples of human communication. The *Ni-Channel* is moderated by volunteers (Kaigo and Watanabe 2007).

The *Ni-Channel* is an anonymous posting system, deviating from the systems of most Internet forums in other nations which often require users to register and email verification to avoid trolling or flaming. Although a name field is available in *Ni-Channel* threads, it is hardly ever actually used for real names. This allows people to post without taking any risk of revealing their identity. It thus creates an atmosphere in which people are able to discuss anything straightforward and fully offensive, mainly due to the fact that nobody knows who is involved in the discussion. The users and fans of the *Ni-Channel* Internet forum cite this as "honesty" and justify the important social role and function the forum serves for Japan (Kaigo and Watanabe 2007).

Highly popularized through this *Ni-Channel* Internet forum, the P2P platform Winny was developed as the successor to a formerly popular P2P software named WinMX (the name Winny is a play on letters of WinMX, as N and Y are the alphabet letters that follow M and X). Winny does not require a central server, and is a "pure" P2P system. Therefore, the network is robust and resilient, making it difficult to terminate. The data communication encoding uses a forwarding function in each computer's cache, in addition to an anonymous bulletin board function, and the design of the system allows each personal user to remain anonymous.

Unfortunately, in late 2005, Winny began gaining much negative media attention and prominence as it became identified as the major vehicle for transferring computer viruses, especially malicious worms like Trojan horses, over the Internet. Malicious worms that include a type of "exposure" virus was transmitted throughout the Winny network to maliciously expose the user's personal chat logs, email data, digital camera pictures, screenshots, password and memos, unknowingly to the user, throughout the network. The worm begins sharing data that it found on the desktop of the user's computer without the knowledge of the user, meaning that by the time the victim realizes his computer has been infected, the situation is disastrous and too late because all the victim's files are already being shared throughout the Internet. The most malicious worms convert the computer into an HTTP server and begin exposing all the data in the computer over the Internet, and all these HTTP servers are linked so they merge all the infected computers together.

The unintentional distribution of sensitive and private information was reported continuously throughout 2005–2006. The confidential customer information owned by various prestigious companies, police records of the Japanese National Police Agency, confidential prison records and judiciary information, and even sensitive information about nuclear power plants and military secrets of the Japanese Ground Self-Defense Force, the Japanese Maritime Self-Defense Force and the Japanese Air Self-Defense Force were all exposed through contagion on the Winny network. One of the most devastating unintentional distribution incidents happened with the controversial *Jyuki-Net* (Basic Residential Registers Network System) access password being exposed onto the Winny network by a civil servant in Hokkaido (Anonymous 2006). Finally, the leaks of military secrets of Japan prompted former Japanese Prime Minister Junichiro Koizumi to plan measures to ban the usage of Winny illegal in Japan to avoid any recurrences of such incidents. The former Chief Cabinet Secretary of Japan at that time, Mr. Shinzo Abe, who later became Prime Minister, held a press conference issuing a warning about using file sharing software such as Winny and advised all Japan's citizens not to use file sharing software, and especially Winny, via the mainstream media. This measure came at a time when sensitive government information at the time was continuously being distributed unintentionally through contagion by Winny. The proposed method to deter further distribution of the information was to have all citizens to stop using Winny and terminating the contagion. The main reason leading to the spread of all these events was through the use of Winny at home, by the employees, civil servants or workers who kept sensitive government information on their personal computers at home. Sometimes, a younger family member would use the Winny network on the same computer of the adult owner without consent, who was using the computer for work. This resulted in the computer being infected and the subsequent distribution of sensitive information by the malicious virus.

The constant media coverage of contagion and Winny made many citizens aware of computer viral contagion, the name of the software and P2P technology. Through the heavy media coverage of Winny, the mainstream media may have once again furthered another element of negative perception of the Internet among the general Japanese public. The individuals or parties that had sensitive information unintentionally distributed by the Winny network were labeled "victims" of cybercrime in the coverage in the Japanese media. Such constant negative coverage may have overly heightened fear of computer viral contagion among the Japanese.

Ni-Channel and Winny had thrust Japan into an unknown territory of lawlessness and this newer media environment was viewed as a frightening space in which anyone might be the next victim. The Winny incidents may have boosted anxiety towards adopting or using the Internet among many members of Japanese society. The mainstream media have repeatedly reported how both the *Ni-Channel* and Winny are problematic to society and are synonyms to address the dark and negative side of the Internet in Japan.

Offensive and Harmful Content, Regulation in Japan and the Internet

Currently, a plethora of potentially socially harmful media content is already available in the normal Japanese market for consumers. Violent or sexually explicit videotapes, books, magazines are available for adults and violent or sexually explicit manga (comic books) are also on sale. Control over distribution of violent content or "content with nudity" is comparatively lenient. What may seem contradictory is that sexually explicit material in Japan was banned until the late 1980s, as images or depictions of frontal nudity were banned as illegal, and so were pictures of pubic hair or graphic depictures of genitals and sex acts (Diamond and Uchiyama 1999). With regard to the greater part of traditional media, the distribution of socially harmful digital content is regulated or has established self-imposed codes in the traditional media sector of Japan. As for regulations, Article 175 of the Japanese Criminal Code prohibits any entity to distribute or exhibit obscene pictures or documents in public, with violation of this code punishable by a prison term or fine.

The broadcast media such as television and radio have a voluntary Broadcasting Ethics and Program Improvement Organization (BPO) that services complaints and discusses problems about broadcast material and sends out advisories to the broadcasters. No legal binding or authority is given to these advisories, but broadcasters are expected to follow and respect any admonitions by these organizations. The Broadcast and Human Rights/Other Related Rights Organization—BRO is the base organization of the BPO that was established by NHK and the National Association of Commercial Broadcasters in Japan.

Along with the BRO and BPO, there is a self-imposed broadcasting code. The broadcasting code is principally targeted at controlling discriminatory or obscene or offensive language to maintain public decency in accordance of the broadcasting laws. The observance of this code is most evident in cases of how foul language in broadcast material is usually omitted through censorship by the broadcaster in the form of a beep sound or other sound effect. Another difference in Japanese media in comparison to US or other European media is how the Japanese media sometimes censor a crime suspect's hands in handcuffs, or hides the face of a suspect (unless the suspect already has a coat or blanket over him or her), to uphold human rights.

The Japanese equivalent of the MPAA—*Eiga Rinri Kanri Iinkai* (the Administration Commission of Motion Picture Code of Ethics) evaluates motion pictures and self-imposed ratings. The ratings are similar to those of the MPAA that administer the following categories: a general audience category (all ages), PG12 (parental guidance under 12), R-15 (restricted under 15) and R-18 (restricted under 18). The discrepancy in age categories with the ratings in comparison with that of the MPAA is assumed to be due to differences in educational systems, where in post-World War II, Japanese junior high school students have graduated at the age of 15 and high school students graduate at the age of 18.

The Computer Entertainment Rating Organization (CERO) conducts rating and evaluation of the Japanese domestic market home electronic or video game machine software and establishes age limit recommendations for consumers. Rating and evaluation by the Japan Amusement Machinery Manufacturers Association (JAMMA) is done for commercial amusement machine game software. CERO is very similar to the Entertainment Software Rating Board of CESA. Although membership is not mandatory for all manufacturers, most game manufacturers and software houses are members of CERO. CERO conducts rating and evaluation by referees that are over 20 years of age and excludes selecting referees that belong to the video game business. The main content of the video game, along with "hidden content" (because video games often have hidden content that can be viewed only when a special code is used) is rated. Since March 2006 in Japan, home video game software rating is currently divided into five rating categories: A—for all ages, B—for over 12 years old, C—for over 15 years old, D—for over 17 years old, Z—for over 18 years old only. The rating is determined at least one month prior to release of the software, and suspension of sales is advised if content deviates from the given standard. Exact rating standard details are not revealed to maintain up-to-date rating standards.

Content descriptor icons are also labeled on the package by the guidance of CERO. These icons may depict the outline of some of the current confidential standards used by CERO: (1) Passionate scenes of sexuality, consanguineous affairs, dating, hugging, kissing; (2) Sexual scenes—costume of high exposure, nudity, lingerie, swimsuits, touching (sexual acts or genitalia is prohibited by CERO); (3) Violence—fights, abuse, torture, and all combative sports such as martial arts; (4) Horror-blood and gore, dead bodies, zombies, ghosts or body part slaughtering; (5) Gambling—illegal gambling activities such as casino games or mah jong, however if no monetary exchange or stripping is involved, the rating becomes suitable for all ages; (6) Crime—murder, robbery, theft, race games and car action violence inclusive of traffic violations, praise or affirmation of criminals; (7) Drinking and Smoking—underage drinking or smoking or unnecessary scenes of smoking or drinking; (8) Drugs—illegal drug usage or dealing; and (9) Language—sexual, discriminating, offensive language (basically adhering to the broadcasting code).

The organization called *Computer Software Rinri Kikou* (Ethics of Computer Software, usually abbreviated *Sofurin*) conducts ethical evaluation of adult-oriented PC games, usually of a pornographic nature. This organization has no legal binding power or any specific authority, and therefore adult PC games can be put on the market without receiving any rating from this organization.

Adult pornographic video material has been evaluated and approved by the *Nihon Bideo Rinri Kyoukai* (Japan Video Ethics Commission) and the Supervisory Committee of Video Ethics, along with the Contents

Soft(ware) Association (CSA) that evaluates video material and adult game software. The main function of this commission is to regulate censorship by prohibiting the display of any graphic sexual pictures and human genitalia.

A noteworthy discrepancy between broadcasting customs of Japan with US or European customs exists in the historically lax attitude of Japan towards broadcasting brief female nudity and fictional violence in contrast to its strict custom and laws prohibiting graphic sex. Depending on the hour of airing, broadcasts that depict brief female nudity may be offensive, but had not been considered a violation of public decency in Japan until recently. Apart from this, one may observe comparatively lower concerns towards fictional violence in Japan—exemplary in Japanese television for children such as animated cartoons. On the other hand, any display of genitalia is in violation of Article 175 of the Japanese Criminal Code.

The main problem with Article 175 of the Japanese Criminal Code becomes evident with the advent of the Internet and new information and communication technologies involved. The law itself is out of date, and is questionable because: (1) the term "pictures" does not technically cover the legal territory of computer "images"; and (2) "obscenity" is not clearly defined enough (Ibusuki 1997). Although strict standards have been imposed on obscene material for traditional media, the Internet environment has introduced conflicting interpretations, therefore, disrupting the legal procedures that are involved in enforcing Article 175.

In this definition, the Internet has an abundance of inappropriate content deemed unfit for the viewing of minors. A major focus of this in Japan is on protecting children from unfit content, and how to shield children from this information, although some of the improper content is usually unworthy for adult users as well. The content in question here is essentially violating online safety standards. Filters with these standards have been quite effective for overseeing web browsing activity or blocking out IP addresses, although even with such filters, web browsing is not 100 percent safe for children.

In the media environment of Japan, broadcast and printed media are allowed to show acted death by actors or animated. Images of death in dramas and such are in abundance in the normal Japanese mainstream media such as television and salient in the *manga* or comic books. However, corpses are usually visually censored in mainstream media broadcasts due to religious and privacy issues. This is not a uniform rule for all nations, so in some broadcasts from other nations, images of corpses from accidents or from wars are televised, and these images become available through the

Internet. In other times, such as during the war in Iraq, hostage executions were being released onto the Internet, and then were uploaded onto file servers (Kaigo and Watanabe 2007). These images are very easy to access, and filters did not stop minors from accessing these images over the Internet.

In a previous study with a colleague, we attempted to examine how Internet and especially *Ni-Channel* users would disseminate information about how to download execution files of a Japanese hostage in Iraq in 2004 (Kaigo and Watanabe 2007). However, instead of these files getting infested all over the Internet, Japanese users repudiated such information like a contagion. The 2Ch users became moderators and began to act as a type of collective pro-social team trying to beat away the information by discouraging any more links. The emergence of a collective ethical "conscience", that was showing up without any governmental bodies making any action illustrates how members of society can act together in cyberspace, and without any management from above, and determine right from wrong.

Unfortunately, this is an anomaly and only because of the extremely violent nature of this specific execution video and mental assimilation among the users due to the fact that the victim was Japanese, the users acted this way. Terrible images and videos are accessed and distributed all the time. The Internet has no regulating body that sets guidelines for what is indecent even for the Internet, so such revolting or shocking images that are socially harmful are easily accessible. The flow of harmful images and videos through the Internet has distorted all trials by the Japanese government and associated organizations to regulate this type of content so that it will not reach the normal user. Another issue is the prevention of cybercrime, as sex crimes and incidents that involve minors accessing Internet friendship (dating) services that have been a persistent problem.

Regional governments have begun taking action by requiring (1) Internet service providers (ISP) to provide filtering software to stop the flow of information that has the potential to harm the minds of youth. (2) Internet service providers must confirm if there are any minors residing with those who wish to join and must inform and easily make available the filters. (3) Facilities that have Internet access must make sure to have youth access devices that have installed filters.

The problem is that filtering software does not solve the problem of minors accessing the horror available through the Internet environment. As switching the URL or using code to overcome keyword filters are easy ways to break filters, new keywords are used to outsmart the filters. The Internet is a dynamic environment, and ordinances will always have the risk of becoming obsolete when dealing with the Internet.

Along with violent and pornographic images, websites that distribute information on how to commit suicide painlessly and bulletin boards that try to recruit people to commit suicide together have also been considered a source of socially harmful content in Japan. These websites and bulletin boards periodically gain mainstream media attention when young teenagers use the websites and commit suicide together. The National Police Agency has asked for cooperation by all Internet- related service organizations to help them save lives when suicide premonitions are made on any of their services.

As a total solution to block harmful information through the Internet is not in sight by technology or regulation, the outcome for the time being now lies in the hands of family ethics and values, because the parent or guardian will need to confront the issue with his or her other family members and begin a discussion. In this case, the family unit will need to decide what is “taboo” and what is “undesirable” and how to deal with such information.

In relation, a well-known phenomenon of antisocial behavior on the Internet such as flaming or trolling has been researched in the past. Some studies have concluded this as an anomaly of the Internet as there are just a handful of people who are naturally trolls, but newer research claims that anyone can become a troll depending on mood and the content of any online discussion (Cheng et al. 2017). Such undesirable behavior seems to be acted out by a small number of individuals in most situations; however, the way the Japanese see this is somewhat unique but similar. In Japan, multiple trolls have repeatedly ganged up on an individual or an organization and the phenomenon is referred to as *Enjyo* which literally means “going up in flames” and is a situation that most Japanese individuals and organizations are afraid of and very much want to avoid. The fear of *Enjyo* can inhibit one from using Twitter as many accounts by individuals have gone up in flames in the past and have been shut down. In severe cases when there have been death threats, the National Police Agency has intervened in past cases. Since 2016, a risk management firm under the Japanese insurance company *Sompo Japan* now offers an *Enjyo* prevention package which provides monitoring, consulting and press conference support. Especially in the case of organizations such as corporations, attacks can continue for a prolonged time and if no measures taken against these attacks, they may even affect stock prices and take an even longer time to recover. If the *Enjyo* is a result of some wrongdoing on the part of an organization, actions such as public apologies are often seen as a remedy to these attacks; however, in the case of individuals, sometimes there are very few ways to have these attacks stop and may require some type of regulation. To illustrate the background of this common anxiety among Japanese organizations, there have been incidents in the past where hotel restaurant personnel began to post private information about celebrity customers in the mainstream media via social media, and, subsequently, this scandalous information made its way back into the mainstream media, such as gossip tabloids and tabloid television. The results were damaging to the organization that was reported in the news and took time for recovery because such information essentially lures trolls and results in creating long-term negative corporate images—in the case of this example, the restaurant and the hotel.

Regulation on part of the governments is demanded by various sectors of society, however, the real issue is about how to create a society that naturally promotes ethics and allows ethical values to permeate into the Internet. Certain facets of Japan will require more regulation or more technological innovations, however, some facets are showing how society can naturally adapt to harmful information when it goes to one extreme, and self-regulate to determine what is acceptable and what is not. The time may be ripe for Japan to confront these issues, instead of hiding them.

Early Social Media in Japan and Aggregators

Along with the early characteristics of Internet platforms that clearly separated the real and virtual world, wariness about entering the Internet environment among a large portion of the Japanese persisted even with the introduction of various social media platforms. Social media connected the detachment of information from people, places and things and along with this new connection of the virtual to the real, the perception of the Internet in Japan also began to change. From 1999, blogs provided a simple interface for publishing content on the Internet and the existence of blogs eventually became well known in Japan from 2003 as they were started by Japanese celebrities and politicians. Unlike other nations, the blogs in Japan that were widespread were not platforms to stand up against the mainstream media, but instead were veritable diaries of people's everyday life, food reviews, or book reviews. Blogs were also seen as useful PR tools and were utilized by businesses for marketing of services and goods. To facilitate this, bloggers were invited to press conferences of announcements of new products or services. However, blogs did not become popular in the same manner as other nations partly because the Japanese Internet environment at the time was still being perceived to be a frightening place to disclose one's private information. In result, information exchange through blogs was not as prevalent among Japanese in comparison to other nations.

In the political sphere, Japanese politicians who were mainly relying on mail magazines to transmit information about their opinions began to shifting to blogs. Their use of blogs had a comparatively short life span because in a matter of few years, the introduction of more popular and widespread use of social networking services gradually led to the use of social media among the politicians. Social media became their preferred method of relaying information to the Japanese public outside the mainstream media. However, the blog interface facilitated the eventual popularization of other social media in Japan such as aggregator sites. These aggregator sites in Japan have become popular tools for gaining information and are a noteworthy facet of Japanese social media.

Famous aggregator sites such as Reddit and Digg are popular among English-speaking Internet users, who wish to gain information efficiently about popular topics that are being discussed in various locations in the Internet that may interest them. In Japan, other than the main popular portals such as Yahoo and Google News, there are similar sites called *Matome Saito* (summary or aggregator sites) that are individually managed blogs that collect and edit information and discussions among Internet users about various popular topics. Similar to the US aggregator sites, the *Matome Saito* in Japan collect information about scandals, pictures, videos, along with news information about politics and society that are popular or relevant to Japanese from all over the Internet. The difference is that the relatively popular *Matome Saito* are blogs maintained by individuals that often scan popular topics in the *Ni-Channel* (*Ni-Channeru* or Ch2) anonymous bulletin board site, and present the information efficiently and concisely so that one can navigate and select information from the colossal amount of information that is being discussed in the various sections of the *Ni-Channel* forum. The information is rapidly and accurately procured, disseminated and delivered to those that view these blogs, often providing news quicker than the main news pages of the popular portals in Japan such as Yahoo or Google, therefore, the *Matome Saito* are considered to be an efficient and reliable source of information, and at times, superior than the portals, even if the content is dubious or outright false from time to time.

The *Matome Saito* also exist as a potential tool for providing important political information and comments about news for those scanning cyberspace who are not traditionally knowledgeable about politics and current affairs. The information in the blogs becomes a convenient guideline for gaining information on current topics; however, many potential problems also exist, because they are the personally edited blogs of entries by anonymous users that do not adhere to any social or ethical norms of journalism. The management of such *Matome Saito* by a single individual every day is close to impossible, so the use of wikis allows for several people to add and renew information on the *Matome Saito*. These individuals can participate in creating and maintaining these sites that require the constant effort of selecting and editing topics from the *Ni-Channel*. Although these *Matome Saito* mainly collect and edit popular information from anonymous bulletin board sites, the information on *Matome Saito* is not always balanced.

The *Matome Saito*, which are prevalent throughout Japanese cyberspace, are popular, and they provide useful sophistry. There are smartphone apps that provide the information on these sites tailored for optimum results on these devices. This allows for many users to conveniently receive common information and political viewpoints via a "meso" medium among popular topics. In result, the *Matome Saito* are creating information sources that provide concisely edited news commentary on various news topics from the *Ni-Channel*.

The *Matome Saito*, or edited *Ni-Channel* thread blogs aggregate information of this largest bulletin board in Japan. The edited *Ni-Channel* thread blogs pick up topics that are popular—particularly hot threads that are getting many entries and viewers. The *Matome Saito,* or edited *Ni-Channel* thread blogs, subsequently pick up the threads that are getting the most attention, or that fall into the category that interests the editor, and will delete the trolling and ASCII art that is rampant in these threads to improve readability, change the color and/or enlarge the font of interesting comments or comments made by the thread originator and upload these as a blog. In this way, the *Matome Saito* is a blog that conveys information in the *Ni-Channel* to a wider audience than the typical *Ni-Channel* user. The blogs are updated daily, and are mediated to a large audience, with the rapid diffusion of hot topics. The blogs edit user-generated content of the *Ni-Channel* into a package with much better readability. Through affiliate programs or AdSense (Google) editors of popular blogs can gain revenue through advertising so the blogs compete with each other to gain more readers by improving readability, add more controversy and alter titles to make the edited threads more attractive. Twitter and other types of social media simply post link to these blogs and proliferate knowledge on these blogs.

The *Ni-Channel-* based content on *Matome Saito* can be divided into news and non-news content. The non-news content, for instance, may be about interesting exchanges between *Ni-Channel* users, material such as stories or personal episodes, or pictures, videos. News content can be from the *Ni-Channel News Sokuhoban* (News flash bulletin board), which is a popular bulletin board. News on this bulletin board can range from stories supplied by news agencies and newspapers to anime or sports. What is noteworthy about the information stemming from this bulletin board is how *Ni-Channel* users will read specific news, and decode the news and comment in a way to re-create the news into *neta* (new material) that can receive more liking by the majority of the *Ni-Channel* users. The exchanges among the users can become new material, or the way the title is presented can lead to entries flooding the thread.

Kashiwabara (2012) illustrated in his study how instead of the traditional Japanese mass media, the *Matome Saito* were central in initially disseminating the November 2010 YouTube video leak of the Chinese fishing vessel ramming into the Japanese Maritime Self-Defense Force vessel the previous month. The videos of the Chinese fishing vessel ramming into the Japanese Maritime Self-Defense Force vessel in October 2010 were not to be made public as the result of a diplomatic agreement between Japan and the People's Republic of China to avert a worsening of relations between the two nations. However, the videos were leaked onto YouTube and, subsequently, the *Ni-Channel* and *Matome Saito* quickly picked up this information and acted as a catalyst to quickly spread the existence of these videos throughout the Internet. As more users became aware of these videos and information became diffused more widely, the popularity of this issue became visible in the rankings of the *Ni-Channel* and in the *Matome Saito*. Such content in the *Ni-Channel* tends to be flooded with comments, or become a *matsuri* ("festival" of flaming) as Tsuda (2009) has described. As the *Matome Saito* were instrumental for diffusion of this news, in result, this incident dramatically increased the perceived reliability and trustworthiness of *Matome Saito*. Among some Japanese, the *Matome Saito* are now considered to have higher credibility than the legacy traditional mass media as a result of the 2010 video incident, because the mass media initially cooperated with government in not showing the footage of these videos and in effect were considered collaborators of the government and also Pro-Chinese and Anti-Japanese for not actively disclosing these videos. The *Matome Saito* do not comply with the government and are not regulated, so information is viewed to be more objective because they do not need to adhere to legal, social or other professional norms or regulations, and supposedly provide both sides of any issue.

A "*Matome Saito no Matome*" is an aggregator of these numerous edited *Ni-Channel Matome Saito* blog pages. The *Matome Saito* aggregator will list either the most recent or most popular blog pages based on *Hatena* bookmarks, which are online bookmarks that are shared through a website, as a social bookmarking service. The *Matome Saito* aggregator will have many blogs, for instance, *MT2* has over 150 registered as *Matome Saito*. The *Matome Saito* aggregator differs from popular aggregators in the U.S. such as Reddit and is also different from Google News in the aspect of having an actual human editor creating the blog instead of having a computer algorithm, which is used by many of the search engine portal news sites.

Hindman (2009) explains how the Internet has made political inequalities already existing more equal, but on the other hand, it has led to newer inequalities. Citizen participation in political voices heard on the Internet may be one of these inequalities, as we can observe in the case of the aggregators in Japan. Hindman (2009) states how linking structures lead to the concentration of certain information that is contrary to the assumption that the Internet will allow for diversification of voices. If only a few visitors arrive at a website, it is narrowcasting while the websites that receive the most visitors continue to expand their viewership/readership. The vastness of available political information and the lack of cognitive resources to process this wealth of information leads people to depend on a limited, handful of convenient sources of political information. In such, Hindman (2009) claims that websites have a tendency to be a "winner takes all", so concentration of online access of audience share due to the nature and structure of the Internet is inevitable, however, Hindman mentions that the Internet has not entirely displaced the legacy media or traditional media.

The case may be similar among younger Japanese that have begun living in a different information environment than the age group above them. The younger Japanese are not so dependent on traditional media when taking into consideration the many different types of information that caters to their needs. The most common way to access news for younger Japanese is through portal sites on mobile devices in recent years (Dentsu Public Relations 2012). Younger Japanese are less dependent on traditional media. In addition, political information in each of the major legacy media outlets has been often viewed to be supporting a specific political view in Japan. In result, many Japanese in their teens and twenties have been reluctant to support those views or subscribe to these legacy media outlets, especially if they have been disinterested or had apathy towards politics in the past and see a grim future that is the fault of the liberal or left-wing policies in control of the government.

The *Ni-Channel* and the *Matome Saito* blogs may have transformed this disinterest towards politics among the youth, by providing additional information or viewpoints similar to their voices on certain political themes. Along with various miscellaneous topics, political news is included in the *Matome Saito* for audiences that have traditionally had no interest at all about politics. Ordinary individuals create the content on *Ni-Channel*, and the *Matome Saito* blogs edit the information to cater to the political inclination of the editor or popular view. The content found on the *Ni-Channel* has an inherent polarizing nature through anonymity and traditionally display direct and hostile interaction among those that make entries (Onishi 2004; Johnston 2006; Fackler 2010; Mie 2013). However, this content style may lead the audience to feel the *Ni-Channel* is more truthful and objective than the traditional media outlets which try to use more indirect expressions when delivering the news (i.e., in news about accidents, a situation in which the victim has had both arms and legs severed may be expressed in the Japanese mass media as "the victim is in a serious condition"). Hindman (2009) has indicated how blogs are now the new elite media, although in the case of the *Matome Saito*, the content is not so much what the blogs originally create, but rather how they edit available user-generated content of the *Ni-Channel* forum that uses politically polarized language. This creates an image of the blogs to be honest political opinions and discussions of the ordinary citizens due to the commentary information presenting both sides of any issue. People browsing the *Matome Saito* sense that they are able to access what ordinary citizens perceive about each issue through the comments, in contrast to reading a column in the newspaper or the opinion of an educated expert on television. Therefore the *Matome Saito* differ from Hindman's view of blogs, in the sense that they are not the original content of a select elite, but they are the popular blogs of people who want to access and read the words of other people on various issues that concern the population at the time.

The *Matome Saito* aggregators then list up the popular blogs, and let people know which topic is "hot" or popular. The problem lies in the fact that the entries of these threads are far from a public sphere. Cass Sunstein (2001) and Yochai Benkler (2006) have pointed out problems in online deliberation. Sunstein (2001) has warned how the Internet will not sustain but rather, fragment a public sphere by providing a chamber that echoes polarization among those in that realm. Benkler (2006) is in accord with the critique of Habermas that suggests commercial mass mediated public discussion is at the level of the "lowest common denominator" and lacks a sense of upholding the public good. The commercially run blogs of the *Ni-Channel* may create a combination of both, but at the same time provide a channel that resembles a trustworthy, objective source of political information. Benkler also has commented on how the Internet lacked the strength to uphold a Jeffersonian model of democracy due to the polarizing views of the users. This weakness may not imply that the Internet is lacking in influencing people who access information. The current *Matome Saito* aggregators may relate to the findings of Yoo (2011), that has stated gatekeepers who do not produce original content can be as influential as the traditional elite media and that the news found in the aggregator's agenda were in sync with the public agenda, suggesting it to have some significant influence over people. Kashiwabara (2012) also probed how the *Ni-Channel* and *Matome Saito* may act as a double gatekeeper when providing information to a wider audience. Aggregators may function as gatekeepers of information, as the legacy media have acted in the past. Williams and Delli Carpini (2000) consider the Internet to have no gates, so there is no such thing as gatekeeping. However, Hindman (2009) claims that there is gatekeeping done through aggregation and that the Internet supports the filtering of the political information. I will now examine how a variation of gatekeeping may be occurring in the Japanese context through reviewing a survey result that probes the *Matome Saito*.

The social media lab of Dentsu Public Relations (the largest media public relations corporation in Japan) named Antenna conducted a survey about *Matome Saito* in June 2012 (random sampled n = 10,000, male = 4359 and female = 5641, age 15 or older, nationwide). Among the 10,000 respondents, 36.5 percent (male = 43.2 percent, female = 31.4 percent) had experience of using or viewing *Matome Saito*.

Among 1200 *Matome Saito* users (male = 600, female = 600), 76.2 percent were using *Matome Saito* more than once a week, and 18.5 percent were heavy users (in other words, these respondents were using *Matome Saito* multiple times per day). Over 30 percent of the male respondents in their teens and twenties, and females in the age range of twenties were heavy users of *Matome Saito*. 90 percent of male respondents in the age range of twenties used *Matome Saito* over once a week. Among males in the age range of 50s and 60s, approximately 10 percent were heavy users and over 70 percent viewed the *Matome Saito* more than once a week. 45 percent of *Matome Saito* users indicated they have seen information initially on *Matome Saito* later being broadcast on television. 14.6 percent of respondents claim they have had this experience numerous times.

The top reason why respondents use or view *Matome Saito* was to "kill time" (48.2 percent), followed by the evaluation that "information is well summarized and easy to view" (38.5 percent), and that they "can access a wider variety of information than mass media" (38.2 percent). Among the respondents in their teens and twenties, the second most common reason was because it is "fun" and among respondents in their 50s and 60s, the second most common reason was because they "can access a wider variety of information than mass media." Therefore, the *Matome Saito* were being considered a form of entertainment among younger users, and as complementary channels of news for the older age group in their 50s and 60s. The younger respondents had a tendency to view *Matome Saito* during transportation (commuting), eating or when in restrooms. One can infer that the *Matome Saito* were being used in brief moments of spare time to browse information.

One explanation for this may be that many *Matome Saito* aggregators are also optimized as smartphone applications such as *2Ch Matome*, *2ChMX* or *MT2*. These applications that can be used on smartphones (multiple platforms) and tablets (multiple platforms) are available as freeware and essentially provide the same information because the logic used for aggregation on these applications is identical.

When a smartphone application user views the content on these aggregator applications, he or she can view both a list of: (1) newly arrived edited thread blog titles; and (2) popular edited thread blog titles of the day. The user can click the thread that seems most interesting and then view the actual linked *Matome Saito* thread. When one views the titles of the threads, one will quickly notice that many titles have multiple consecutive letters of "w" after them, which are the equivalent of "lol" because "w" refers to the Japanese word *warai* or laugh. The term *omoshiroi* in Japanese used to explain "interesting", also has the meaning "fun" and may mean both in certain contexts. These threads are usually ones that fall into the category of "material" that has gone through a "festival" of *Enjyo* or flaming. The edited threads usually provide a cleaned up insight into what people (users of *Ni-channel*) think/thought about this particular issue. The user can check to see if their opinion is the same with the majority of entries of the edited blogs, and he or she can also view how their opinion is viewed among others if they are able to find comments towards someone with the same opinion. Many *Matome Saito* aggregator users do not view the *Ni-channel* forum, and rarely check the actual source, if ever, so the aggregated information source becomes their sole source and may become the source of conversation topics among friends that may or may not use the aggregators.

The popularity of *Matome Saito* has verified it to be an important information source in a positive light. They are allowing many Internet users to conveniently receive common information and political viewpoints via a "meso" medium among popular topics through smartphone apps that provide the information on these sites tailored for optimum results on these devices. They are creating interest in politics among the youth that use this medium for news and information. Most of the thread titles are sensational, as the *Matome Saito* managers need to create revenue by having as many viewers to their blogs as possible, so the current flow of information may seem alarming but the actual effects on the general public of Japan of such, sometimes xenophobic and aggressive arguments in these threads need further research.

References

Anderson, J. R. (1980). *Cognitive psychology and its implications.* New York: W.H. Freeman.

Anonymous. (2006, 29 March). *Hokkaido de Winny niyoru jyukinetto no jouhou rouei.* [Jyuki net information leaks by Winny at Hokkaido.] Kyodo News.

Bandura, A. (1995). Exercise of personal and collective efficacy in changing societies. In A. Bandura (Ed.), *Self-efficacy in changing societies* (pp. 1–45). Cambridge: Cambridge University Press.

Bandura, A. (1997). *Self-efficacy: The exercise of control.* New York: W.H. Freeman.

Barker, L., & Edwards, R. (1980). *Intrapersonal communication.* Dubuque: Gorsuch Scarisbrick Publishers.

Benkler, Y. (2006). *The wealth of networks: How social production transforms markets and freedom.* New Haven: Yale University Press.

Bush, T. (1995). Gender differences in self-efficacy and attitude towards computers. *Journal of Educational Computing Research, 12*(2), 147–158.

Cheng, J., Bernstein, M., Danescu-Niculescu-Mizil, C., & Leskovec, J. (2017, February 25–March 1). *Anyone can become a troll: Causes of trolling behavior in online discussions.* Paper presented at the 20th ACM Conference on Computer-Supported Cooperative Work and Social Computing, Portland.

Chua, S. L., Chen, D. T., & Wong, A. F. L. (1999). Computer anxiety and its correlates: A meta-analysis. *Computers in Human Behavior, 15*(5), 609–623.

Day, D., & Makirinne-Crofts, P. (1997). Computer anxiety and the human-computer interface. *Proceedings of the HCI'97 conference on people and computers XII,* 85–100.

Dentsu Public Relations. (2012, July 18). *Houdou Shiryo: Atarashi Jyohou Ryutsu Kouzou—Himatsubushi Media Matome Saito no Riyou Jittai Chousa.* [Press Release: New Information Distribution Structure: Media to Kill Time Matome Saito Usage Survey]. Retrieved from http://www.dentsu-pr.co.jp/wp-content/themes/dpr_themes/release/20120718_report.pdf

Diamond, M., & Uchiyama, A. (1999). Pornography, rape and sex crimes in Japan. *International Journal of Law and Psychiatry, 22*(1), 1–22.

Dyck, J. L., & Smither, J. A. (1992). Computer anxiety and the older adult: Relationships with computer experience, gender, education and age. *Proceedings of the Human Factors Society 36th Annual Meeting, 1,* 185–189.

Eastin, M. S., & LaRose, R. (2000). Internet self-efficacy and the psychology of the digital divide. *Journal of Computer-Mediated Communication, 6*(1), 0. doi:10.1111/j.1083-6101.2000.tb00110.x.

Fackler, M. (2010, August 28). New Dissent in Japan is Loudly Anti-Foreign. *New York Times.* Retrieved from http://www.nytimes.com/2010/08/29/world/asia/29japan.html?pagewanted=all&_r=0

Gilroy, F. D., & Desai, H. B. (1986). Computer anxiety: Sex, race and age. *International Journal of Man-Machine Studies, 25*(6), 711–719.

Hindman, M. (2009). *The myth of digital democracy.* Princeton: Princeton University Press.

Hoffman, D. L., Novak, T. P., & Schlosser, A. (2000). The evolution of the digital divide: How gaps in Internet access may impact electronic commerce. *Journal of Computer-Mediated Communication, 5*(3), 0. doi:10.1111/j.1083-6101.2000.tb00341.x.

Ibusuki, M. (1997). Legal aspects of cyber-porn in Japan. *Lex Electronica, 3*(1), A.3.

Igbaria, M., & Chakrabarti, A. (1990). Computer anxiety and attitudes towards microcomputer use. *Behaviour and Information Technology, 9*(3), 229–241.

Johnston, E. (2006, March 14). Net boards venue for faceless rightists. *Japan Times.* http://www.japantimes.co.jp/news/2006/03/14/news/net-boards-venue-for-faceless-rightists/#.Ui1gObxPpkA

Kaigo, M., & Sasaki, T. (2001 November. 17–19). *Cognitive and affective factors of new information and communication technology usage and the digital divide in Japan.* Paper presented at IAMCR/ICA Joint Symposium on the Digital Divide, Austin.

Kaigo, M., & Watanabe, I. (2007). Ethos in chaos? Reaction to video files depicting socially harmful images in the channel 2 Japanese Internet forum. *Journal of Computer-Mediated Communication, 12*(4), 1248–1268.

Kashiwabara, T. (2012). A consideration of news communication by "2-channel thread editing blog". *Mita Tetsugakkai, 128* (2012. 3), 207–234.

Mie, A. (2013, January 8). Xenophobia finds fertile soil in web anonymity. *The Japan Times.* Retrieved from http://www.japantimes.co.jp/news/2013/01/08/reference/xenophobia-finds-fertile-soil-in-web-anonymity/#.Ui1atLxPpkA

Newhagen, J. E., & Bucy, E. P. (2004). Routes to media access. In E. P. Bucy & J. E. Newhagen (Eds.), *Media access: Social and psychological dimensions of new technology use* (pp. 3–26). Mahwah: Lawrence Erlbaum Associates.

Nishigaki, T., & Lewis, J. (2001). *Internet de nihongo wa dounaru. [What will happen to the Japanese language because of the Internet].* Tokyo: Iwanami Shoten.

Norman, D. A. (1981). *Perspectives on cognitive science.* Norwood: Ablex.

Onishi, N. (2004, May 9). Japanese find a forum to vent most-secret feelings. *The New York Times.* Retrieved from http://query.nytimes.com/gst/fullpage.html?sec=technology&res=9F01E5DE153CF93AA35756C0A9629C8B63

Parasuraman, S., & Igbaria, M. (1990). An examination of gender differences in the determinants of computer anxiety and attitudes toward microcomputers among managers. *International Journal of Man-Machine Studies, 32*(3), 327–340.

Paxton, A. L., & Turner, E. J. (1984). The application of human factors to the needs of the novice computer user. *International Journal of Man-Machine Studies, 20*(2), 137–156.

Rogers, E. M. (1986). *Communication technology.* New York: Free Press.

Simon, H. A. (1979). Information processing models of cognition. *Annual Review of Psychology, 30*, 363–396.

Sunstein, C. (2001). *Republic.com.* Princeton: Princeton University Press.

Szajna, B., & Mackay, J. M. (1995). Predictors of learning performance in a computer-user training environment: A path-analytic study. *International Journal of Human-Computer Interaction, 7*(2), 167–185.

Ting-Toomey, S. (1988). Intercultural conflict styles: A face negotiation theory. In Y. Y. Kim & W. Gudykunst (Eds.), *Theories in intercultural communication* (pp. 213–235). Nevbury Park: Sage.

Tsuda, D. (2009). *Twitter Shakairon Aratana Real Time Webno Choryu [Twitter social theory—A new real time web tide].* Tokyo: Yousensha.

Van Dijk, J. (2004). Divides in succession. Possession, skills and use of the new media for participation. In E. P. Bucy & J. E. Newhagen (Eds.), *Media access: Social and psychological dimensions of new technology use* (pp. 233–254). Mahwah: Lawrence Erlbaum Associates.

Van Dijk, J., & Hacker, K. (2003). The digital divide as a complex and dynamic phenomenon. *The Information Society, 19*(4), 315–326. doi:10.1080/01972240309487.

Weil, M. M., & Rosen, L. D. (1995). The psychological impact of technology from a global perspective: A study of technological sophistication and technophobia in university students from twenty-three countries. *Computers in Human Behavior, 11*(1), 95–133.

Whitley, B. E. (1997). Gender differences in computer-related attitudes and behavior: A meta-analysis. *Computers in Human Behavior, 13*(1), 1–22.

Williams, B. A., & Delli Carpini, M. X. (2000). Unchained reaction: The collapse of media gatekeeping and the Clinton-Lewinsky scandal. *Journalism, 1*, 61–85.

Yoo, S. W. (2011, May 25). Google and Yahoo as the Next Gatekeeper. *Examining Agenda Setting Effects of News Aggregators.* Paper presented at the annual meeting of the International Communication Association, TBA, Boston. Retrieved from http://www.allacademic.com/meta/p492217_index.html

Yotsumoto, M. (2002). I-Life: Jouhoukashakaini ikiru-broadband & ubiquitous computing no kandokoro—[I-life: Life in the information society. Points in broadband & ubiquitous computing.] *Dai4kai seikatusha jouhouriyou chousa report 2002.* [2002 4th Report on findings of the survey of information use among consumers.] Dentsu Institute of Human Studies.

Muneo Kaigo is an associate professor in the Faculty of Humanities and Social Sciences at the University of Tsukuba in Japan. He teaches courses in media communication in the Graduate School of Humanities and Social Sciences, media management in the Graduate School of Business Sciences and communication science in the College of Comparative Culture at the University of Tsukuba. He is currently leading a joint research project on social media uses among local municipalities in Japan with cooperation from the municipal government of Tsukuba and Intel Corporation Japan.

CHAPTER 2

Social Media in Japan and the Great Eastern Japan Earthquake

Muneo Kaigo

Web 2.0 and Social Media in Japan

The transition to Web 2.0 has allowed many individuals to easily upload and share information on the Internet. Japanese Web 2.0 services still continue to roll out in Japan, and there have been numerous social media that have been developed in Japan for the benefit of the Japanese population. For instance, in 2005, YouTube was launched as a video sharing site and soon after, the Japanese video sharing service, Nico Nico Douga was experimentally introduced in 2006. The most notable characteristic of Nico Nico Douga is that all users can add comments superimposed directly onto the shared video image. Nico Nico Douga started its service in December 2006 and grew rapidly during 2007. Spikes in access during peak hours (from 7 p.m. to 2 a.m.), weekend and holidays required limits to be placed on the number of users that can access the service so as to avoid overloading the servers. This resulted in the introduction of a premium membership service that allowed unlimited access during these times. Nico Nico Douga originally started by accessing and retrieving videos on YouTube; however, in February 2007 YouTube servers

M. Kaigo (✉)
Faculty of Humanities and Social Sciences, University of Tsukuba,
Tsukuba, Ibaraki, Japan

M. Kaigo (ed.), *Social Media and Civil Society in Japan*,
DOI 10.1007/978-981-10-5095-4_2

terminated all access from Nico Nico Douga in order to stop Nico Nico Douga from free-riding. This resulted in the creation of a new service that allowed several different kinds of video codecs to be uploaded. In 2007, Ustream was launched and this provided live video streaming, and in the same year Nico Nico Namahousou (live broadcast) was introduced in Japan, with the press conferences of politicians becoming available through this platform. During this same period, a number of social networking services were also introduced into Japan.

The definitions and history of social networking services have been well documented and there have been various comparisons of platforms in different contexts (boyd and Ellison 2007; Raacke and Bonds-Raacke 2008). The social networking services of Japan have been evolving in a different context while numerous Japanese companies attempted to popularize their own individual platforms. Different platforms have been emerging from Japan and the initial spread of social networking services that took off in Japan was among these Japan-born platforms, which were usually attempting to catch up with other services that had been launched earlier in the United States and other areas. Following the introduction of social networking services such as Friendster and MySpace, Japanese social networking services such as Mixi and GREE were introduced in 2004. Becoming a member of some social networking services of the "Made in Japan" platforms required invitations from others who were already members. The existence of such a formal 'invitation-only' membership system gave a sense of security to the many Japanese users who became new members of these services. Compared with the unprotected and threatening environment of the blogs in Japan (see Chap. 1), Mixi was regarded as a safe area to exchange information and ideas, so for a period of time, it became the most popular social networking service in Japan. The 'invitation-only' system of participation made people feel more at ease, and the "footsteps" function that recorded and showed you a list of people who had viewed your Mixi site, added to the sense of security. Mixi also had a community function, so that its members were able to connect with each other based on their specific interests or hobbies, local hometowns or schools that they graduated from. Another positive characteristic of Mixi for Japan, in addition to its emphasis on protecting privacy, was in the way it was rolled out as a mobile phone-oriented service. The service continued to grow quite rapidly and became, for a time, even more popular than the *Ni-Channel* bulletin board (see Chap. 1). The web-enabled mobile phone interface of Mixi allowed for text input to be very similar to text messaging which facilitated the increase of younger users because a large population of the youth in Japan had relatively few computers in the home and/or an Internet connection. As many users were able to view Mixi through their web-enabled cellular phones during their daily commute, they could use their cameras on their phones to post pictures with an easy email interface, and users continued to increase until Twitter and Facebook eventually overtook Mixi and GREE following the Great East Japan Earthquake.

According to an ethnographic study of users of Mixi and MySpace, Takahashi (2010) pointed out that Japanese engagement in the social context with social networking services had the following dimensions: (1) information-seeking activity; (2) connectivity; (3) bricolage; and (4) participation. Takahashi states that Japanese SNS use has more of a collective dynamic (with a focus on "us"), in comparison to social networking sites initially developed in the United States such as MySpace, that is more about the individual user (oneself or "me"). Takahashi also explains the constructs of *uchi* (inner) and *soto* (outer) to explain how users attribute individual properties to their online personas in different platforms while often simultaneously using pseudonyms. The web interface of Mixi also creates a boundary that reflects society, and this interrelates with preexisting cultural behavior and distinctions between ethnic groups online.

Unfortunately, Mixi and MySpace are no longer the dominant SNS in Japanese and US cyberspace. According to a report published by Nielsen Netview of Japan, in March 2013, Facebook had the most unique visitors among the various SNS in Japan, with a total of 17,515,000. In contrast, Mixi, formerly the most dominant social network in Japan, had only 4,468,000 visitors in the same month. For the purposes of comparison, data from July 2011 report 14,033,000 visitors to Mixi and 9,504,000 visitors to Facebook, showing that within two years, the SNS usage ratio in Japan had changed dramatically. In less than two years, the number of users of each service changed rapidly, resulting in Facebook being the most popular SNS in Japan by March 2013. One possible factor in the rapid decline of Mixi may have been its relative lateness in adopting a smartphone interface for the service. The number of Mixi and GREE users in Japan continue to decline and they have completely lost their popularity to Twitter and Facebook in Japan; however, both services still exist and they have shifted their businesses more towards the areas of social gaming and the gaming community. What then is the most popular social media in Japan at the time of writing? The answer is LINE.

In 2009, the WhatsApp voice, video calling and messenger service, was launched but adoption among Japanese was sluggish at best. In 2011, after the Great East Japan Earthquake, the similar but different messenger service known as LINE initiated its development and launching in the same year due to the high perceived demand for being connected to family and friends during this disaster. The underlying notion of communication through ICT over the Internet as asynchronous became obsolete with these voice, video calling and messenger services because direct communication among users in close to real time over the Internet became increasingly prevalent. The combination of the increase in smartphone use, the Great East Japan Earthquake and LINE all acted as catalysts to push the Japanese to overcome their previous uneasiness towards the Internet and ICT (See Chap. 1) and led the way for many Japanese to go online. Currently, LINE is the most popular social media platform in Japan; however, its use among most Japanese is mainly as a person-to-person or group messaging service that is more convenient than the mobile carrier messaging services because it is not carrier dependent. Although LINE is popular in Japan, the most common usage of LINE is for private messaging, as it provides a very convenient method to communicate between close friends and family, especially during a crisis like the Great East Japan Earthquake. However, it is different from the dynamics of social media such as public sharing information to your community or a larger audience in Japan. Therefore, this book will focus more on the types of social media usage that are not limited to private messaging uses.

The Great East Japan Earthquake: A Recount of Events in Tsukuba, Ibaraki

March 11, 2011, 14:46

The prefecture in Japan in which I have been living, Ibaraki had been experiencing minor earthquakes prior to March 11, 2011. A major tremor two days ago, on March 9, 2011, was treated as another earthquake. For most of the inhabitants, the earthquakes were becoming another routine nuisance of everyday life. At 14:46, I had returned to our living quarters to pick up some documents, when I began to hear a hum and started feeling small shakes. The shakes continued to gain in strength and as if some turbo-like mechanism kicked in, everything in the room I was in began falling off the shelves. Things started violently flying onto the floor. The lights suddenly all went off, which meant that I was experiencing a total power outage which is very rare in Japan, so this was a cue to tell me that what I was experiencing was no ordinary earthquake. As everything that can possibly fall out of our cupboards was falling onto the floor, large shakes continued and then it finally stopped. I looked around our flat and quickly checked what was damaged and evacuated our flat.

March 11, 2011, 14:50

Many other people living in our neighborhood had evacuated their homes as well. They joined each other around the bench I had sat on. Mothers were trying to contact children on their mobile phones, and wives were trying to contact husbands at work; however, no one was able to get through. As a result of everyone trying to make contact at the same time, no mobile voice calls could get through on any of the major mobile carriers. I tried accessing the Internet, but was unsuccessful as everyone else was also having trouble getting through to any mobile Internet access. Even the mobile carriers' messaging services were congested at the moment right after the first quake because everyone in the area was trying to contact everyone else at the same time.

As all the stereo systems and televisions that we owned needed an electrical outlet, we discovered that most of us did not have any battery-operated portable media to gain access to information of what was happening due to the mobile phones not being able to receive any data. The continuous shaking left us in awe as we watched our houses be rattled by the subsequent shakes. We had been shut out from any media or information for about 20 minutes due to the power outage and congestion on the mobile phones, so I next walked over to my parking lot and decided to get some information through my car's radio and television. Twenty minutes after the first shake, I very quickly learned through the television in my car that this was a very severe, big earthquake, but none of the radio stations nor the television stations I accessed knew exactly what was going on at that moment. All they were repeating was that the epicenter of the first quake was very close to the one that we had two days ago, on March 9, 2011, but much bigger.

March 11, 2011, 15:20

A person living in our neighborhood sitting on the bench had found and brought along a portable radio. As a result, we started receiving real-time reports of the earthquake and about the aftershocks that had been continuing. The radio announcer mentioned that the initial earthquake in our area had a seismic intensity scale of four (to which everyone listening in my vicinity immediately disagreed) that was later upgraded from four to six. We also started receiving warnings of tsunamis of over ten meters along the northeastern sea line of Japan. My wife and I started sending out SMSs to our families in an attempt to communicate with them. The SMSs sent out at 15:24 were properly received, but with a time lag of over an hour as we later learned.

March 11, 2011, 16:00–17:00

As the earth continued to shake every few minutes, I went back into our flat and confirmed that the electricity, gas and water were all unavailable. Water was coming out of some of the faucets in the nearby park and the public washroom was also operating. However, later that day, this small supply of water also came to an end. My wife and I took refuge in our car because it was equipped with a heater, along with a radio and television. The Tsukuba municipal government was providing shelter at a local elementary school, something which was being announced by a loudspeaker on top of a car driving around the neighborhood as our municipality does not carry a city-wide speaker system. Due to the lack of earthquake fortification of this magnitude, every time it shook, they were having the people inside evacuated outside of the school. This meant that people had to leave the building every few minutes. We later learned that many residents were also not able to hear what was being said on the loudspeaker due to the direction of the speaker or the speed of the car moving around affecting reception of the sound to be very bad for most people.

March 11, 2011, 18:00

The Internet connection on my smartphone returned and was now functional. As my computer had become unusable after the initial earthquake, my smartphone became my main communication device. As we watched television coverage or listened to the radio in my car and starting getting some of the visual coverage of horrifying destruction, I began sending out emails to see if our Internet connection was truly functional. After receiving messages through email, I began sending out emails and text messages and starting posting my status on Facebook. I received many comments and messages, asking me how I was, telling me others were doing well too. Encouraged by the good 3G Internet connection, I tried to Skype Out to my mother living in Tokyo who I had not been able to contact since the largest earthquake due to the mobile voice lines being overloaded. To my surprise, voice over IP through Skype worked without many flaws.

March 11, 2011, 20:00

Mobile congestion was slowly improving; however. considering how public transportation was being disrupted in Tokyo, it still seemed difficult to get a normal voice line to anyone. My wife sent a message to me through our mobile phone carrier to experiment and see whether or not the messaging system was working. I was able to receive her message with a time lag of about ten minutes by this time.

March 12, 2011, 04:00

Electricity came back on around 4 a.m., and soon we were able to get information about the other affected areas through our television set. As our cable TV connection seemed to be functioning, I quickly scavenged the mess in our flat caused by the earthquake and reconnected our cable modem and WiFi and had our Internet connection back online. Later that morning I accessed the Tsukuba city website; however, very little information was available. Fortunately, to my surprise the Twitter account was feeding information continuously at that time, so to our relief, information continued to come out of the Twitter account of the Tsukuba municipal government. The local community FM radio station was also transmitting local emergency information. The water system was down, so information on where to receive emergency water supplies was being announced through these channels along with cars with the mounted speakers. Television and radio stations were still reporting about the destruction in the northeastern area of Japan, so very little relevant information about our area was available on the major media outlets. Voice communications were still disrupted, although the Internet had now become essentially functional…

In this recount of the Great East Japan Earthquake, in the context of media access in Tsukuba, Ibaraki, one can observe how in the event of failure of lifelines due to a severe natural disaster, (1) voice communication through normal telephone lines becomes difficult, however, Internet access through mobile devices is relatively robust and resilient in comparison to normal telecommunication channels, therefore VoIP (Voice over IP), (2) SMS and email were quicker solutions for communication and overcoming telecommunication breakdowns in disaster situations. (3) In contrast to websites, which required editing with computers, social media is more effective in disseminating information on the Internet. For example, SNSs like Facebook are convenient for efficiently communicating with friends and family, and micro-blogs like Twitter can widely disseminate information in short text messages, and (4) radio was effective because of the portability of the device required for receiving radio waves, in comparison to televisions are usually dependent on an electricity outlet for it to normally function.

During the Great East Japan Earthquake, Twitter was a valuable communication channel in Tsukuba, as will be considered in detail later in this chapter. Prior to that analysis, I shall first examine relevant perspectives in relation to social media and subsequently social capital in relation to social media.

Twitter in Japan, Social Capital and the Great East Japan Earthquake

The Great East Japan Earthquake affected a large proportion of Japan's population and its effect had many dimensions, but two major aspects stand out which were devastating as they destroyed the lives of many. The horrifying spectacle of the actual shaking from the earthquake and tsunami that continued to swallow houses and people that dominated the mass media is the infamous and very well-remembered dimension. The other well-known dimension that stands out is the nuclear accident, evacuation and accompanying following problems that stunned Japan. The minor dimensions that affected Japan are how a large number of people were left stranded without public transportation to get home, and the interruption of communication services through congestion and/or basic lifeline infrastructure such as electricity, water and gas.

During and after the Great East Japan Earthquake and the subsequent tsunami, many areas of the northeastern coast of Honshu experienced devastation beyond the range of most of the contingency plans that had been put in place to protect people in such a disaster. As the large-scale disaster exceeded the emergency planning of the Japanese ministries that were responsible they were still coping; however, the nuclear disaster that followed left Japan in a state of total disarray. The nuclear accident and meltdown of the Fukushima Daiichi nuclear power plant overwhelmed the public and the government during the explosions, and, to this day, continues to plague Japan with its numerous and various long-lasting effects of radioactive contamination during and after the Great East Japan Earthquake.

During such devastation, social media cannot be used in many of these areas because large sections of the infrastructure necessary for communication through normal means have obviously been destroyed. Electricity supply becomes impossible as cables are broken. Communication antennas are damaged. Indeed, even if the antennas are not damaged, emergency batteries eventually die out due to the lack of electricity because of the prolonged shortage of supply. In the case of the Great East Japan Earthquake, cables for international communications were also damaged. In such a case of large-scale destruction, mobile phones cannot be used for communication, although at the time of the disaster, most Japanese were heavily dependent on mobile phones and carrier messaging services as these were regarded as the most reliable medium for communication. For the areas in northeastern Japan that had the most destruction after the earthquake, computers and other information and communication technology becomes a low priority as people have other, more urgent needs such as heat and other essentials. The necessity of heat to keep warm is of utmost importance, or the need of food to eat, water to drink. The need for real, actual people that can help them becomes a much higher priority than the need for interacting with others through social media. A large-scale disaster with a wide range of destruction can nullify any argument of how important social networking services are, or how people can utilize social media. Human support from first responders, rescue personnel and machines such as automobiles or helicopters for evacuation or other purposes are obviously a higher priority than being online in such severe situations.

What use is social media, then? Unless everything is destroyed, then some or large parts of the infrastructure required for communications may still remain functioning. Even in the areas heavily affected by the Great East Japan Earthquake, if this is the case, then communication through email, Twitter and Facebook can function even if mobile carrier messaging becomes congested. During this disaster, the case of Tsukuba is an example of how social media such as Twitter can allow municipalities with mid-level disruptions to communicate to citizens and allow them to "walk a tightrope" situation until normal telecommunications are back.

Although all social media at the time functioned in various beneficial ways, I will discuss the details about the problems and possibilities of Twitter use and, in particular, how it functioned during the disaster because the use of the platform in Japan is potentially allowing for new contexts of communication that were impossible before. Twitter is basically a beneficial media to use in the following three general methods or contexts: (1) it allows for people to easily create new personal networks and receive exposure to information that could not be available prior to the existence of this platform; (2) it allows for a different channel for everyday online communication with acquaintances; and (3) it allows for simpler information transmission on the

Internet. Following someone on Twitter allows for the follower to know what the account holder is thinking about, and even though thoughts are being transmitted as tweets, the follower feels these thoughts to be more real, building the motivation for followers to follow someone's tweets on Twitter. The act of following the tweets of famous people like actors, artists or famous writers allows ordinary users to read what normally out-of-reach famous people are thinking about at that moment indirectly (although sometimes the accounts of famous people directly communicate to ordinary citizens as well). Following a friend's friend on Twitter may start out as a virtual friendship, but may evolve more easily into a real-life friendship if there should be an occasion to actually meet. In such ways, Twitter allows for people to create new personal networks and receive exposure to information that was not available to them before.

Communicating with people on Twitter that you interact with daily in real life is a popular means of communication among Twitter users in Japan, as a more indirect method of interaction. For example, a phone call or email is a direct person-to-person communication mode that requires a reply or a formation of synchronous communication whereas a tweet is a more indirect mode because it is directed from one person to many people. In other words, a tweet is a message that may be intended for a certain person, however a reciprocal person-to-person communication or interaction is not always being demanded in this situation. A tweet does not require a reply, so interaction becomes in contrast, less obtrusive, voluntary or informal for Japanese people. Communication is also effective and efficient, because communication is one person to many people. The follower of this type of a tweet has the freedom to choose when and whether or not to respond, so there is less pressure for both sides in comparison to normal one-to-one interaction, such as an email or a phone call.

Opening a Twitter account and tweeting are both simple and the current limit of 140 characters does not require a lot of information to be transmitted in each tweet. This is actually a benefit for a wider range of people, because it allows account holders with little literary talent to express their thoughts or ideas. The 140-character limit also applies to Japanese; however, as Japanese characters also use *Kanji* (Chinese characters), a lot of meaning can be condensed through one or a couple of characters and this allows for a lot of meaning to be included in a single tweet. The 140-character limit is much more restrictive if using an alphabet-based language, but because the Japanese written language can condense a lot of information even with this constraint, this actually promotes messages to be created in a fashion to be more simple and easy to understand, but not necessarily truncated or short. Twitter is also friendly for the web-enabled mobile phone and smartphone user. Account holders that do not have a lot of time, nor own a computer can still have access to Twitter and communicate. The properties of the Japanese language and characters and smartphone and web-enabled phone usability may account for the steady popularity of Twitter in Japan.

In contrast to all the preceding possibilities of Twitter, some problems exist among Japanese Twitter users. It can lead to: (1) a weakening of personal relationships; and (2) difficulty in verifying credibility of information. As one is added as a follower to an account, two users are connected virtually. However, that may be the conclusion to this relationship. In other words, the two users are only connected by a virtual link, and no other activity will ever happen. Some text messages may be exchanged, but that may not contribute to a real relationship. For a solid relationship, communication and empathy is required. Such a diluted, weak network may give false images or may lead to various misinterpretations and problems. Verifying the credibility of information is also a problem with Twitter. False rumors or false stories can be communicated very rapidly by retweeting other tweets without verification of the original information source. Retweets may be recreated, unaltered to followers, or be manually altered by adding RT in front of the original tweet. In either case, rumors or stories can spread at rapid pace, and may be dangerous if false or damaging.

Through reviewing the pros and cons of Twitter usage, one can observe both the various possibilities and also the problems of this platform. It can be used for beneficially supporting everyday life and communications; however, it can also be latently problematic in some other instances. As Internet and social media usage is becoming incorporated into the daily lives of most people in Japan, more time is spent on communication through these methods. Even with potential problems, the prevalent usage of social media such as Twitter allow for new forms of communication that are proving to be more beneficial than problematic, in various ways that were unimaginable before. One aggregate benefit of social media that we observed through Twitter use, especially during the Great East Japan Earthquake, can be the formation of social capital.

Social capital is a well-known construct, but it is one that is capable of a variety of different interpretations among many social scientists in different fields, and with those various interpretations, becomes quite complex. Bourdieu (1986) defined social capital as the collective resources or assets that one is able to maintain as a consequence of the activities of social connections with other people. Through the better connection among people or to groups, social capital is enhanced and develops. Bourdieu (1986), pointed out three forms of capital—economic capital, cultural capital and social capital. Cohen and Prusak (2001) described social capital as the active connections among people bound by trust, mutual understanding and values to make cooperative activities properly function.

Social capital can mobilize resources of an individual and lead to function for the betterment of a community (Helliwell and Putnam 2004; Granovetter 1973). Putnam connected social capital to the importance of civic engagement and voluntary organizations for political participation. Social capital is both a structural phenomenon (social networks) and a cultural phenomenon (social norms) (Norris 2001). According to Putnam, the networks of civil society, friends, colleagues and neighbors, can build the conditions for collaboration, coordination and cooperation to generate some type of common good. So if the social bonds are denser and richer, the social infrastructure becomes more vibrant and solid. This definition has three important elements: networks, norms and trust. Putnam believed that dense networks will cultivate more social capital and can build solidarity among other networks. The norms are the mutually shared reciprocal correspondence among members of the community. Trust means having faith in other people, even if one is not familiar with the other. The consolidation of networks, norms and trust are interconnected and mutual. Social network and social support can be considered important pillars of social capital. Trust is incorporated in social support and social networks and this creates a sense of community.

There are a number of studies that employ the ideas of bonding and bridging with regard to social capital. Bonding in social capital refers to the strong ties or close-knit relationships which are founded on similarity, intimacy and frequent contact (Williams 2006; Putnam 2000; Granovetter 1973), such as those found between family members and close friends. Bridging in social capital are the connections among various networks or set of contacts and the weak ties are the ones that appear when people from diverse backgrounds create relationships with each other (Putnam 2000; Granovetter 1973). Adkins (2009) mentions how the networks of weak links are less likely to provide strong support, but the bridging of relationships are a positive feature as this allows for exposure to different people. The trust between nodes or people and linking to new or different resources that were not available through the strongly bonded connections are similar to what Granovetter (1973) called the strength of weak ties.

Putnam has lamented how there has been a decrease in participation in regards to civic activities in the United States and provided the evidence of a decline in America's social capital (Putnam 2000). Putnam pointed out that the increased use of electronic media for entertainment and communication is one of the reasons leading to the decline of social capital. However in the year 2000, Putnam was not able to mention that the Internet is also a definite contributor to the decline of social capital, and that has led the pathway for many to claim that online activities on the Internet may serve as a potential building block to rebuild declining social capital in many of the developed nations. Online interaction and social exchange among individuals and groups can create social capital, through linking people together. One can suggest that the Internet is an interactive means for connecting people and can have a positive role in constructing social capital.

According to Quan-Haase and Wellman (2005), social contact and civic engagement are two complementary uses of social capital. With inexpensive and convenient ways of communication available now through the Internet, remote and distant communities and areas can now be easily reached and the Internet can provide an arena for social capital to provide existing social ties and relationships among people and friends (Chen et al. 2002). The Internet has provided a space for the growth of online communities and social networking services. One of the main reasons for people to join social networking services is to keep in touch with old friends and meet new ones (Joinson 2008) as individuals are able to maintain and build their social networks (Ellison et al. 2007). Some studies have provided some evidence that Internet use can enhance civic and political participation among those online (Valenzuela et al. 2009; Howard and Gilbert 2008; Scheufele and Shah 2000). People now use the Internet to gain support (Bargh and McKenna 2004), link up with charitable institutions (Raine et al. 2005) or engage in political deliberation (Smith and Raine 2008). Positive results between Computer-Mediated Communication (CMC) and community has been documented in the past (e.g. Wellman and Gulia 1999; Schuler 1996; Baym 1995). Hence it is easy to infer that social media also has the potential to be regarded as a possible space to connect and build a community.

In studies of social capital in relation to the network society, the majority of the results are claiming that the Internet encourages social connections and involvement rather than prior concerns of decrease of social involvement or social displacement (Valkenburg and Jochen 2007). The case of social media is the same, as studies are indicating that social capital is greater among those that actively use SNSs (Ellison et al. 2011; Pfeil et al. 2009; Steinfield et al. 2008). Kobayashi and Ikeda (2005) have indicated that trust and reciprocity are such observed elements of social capital during ICT usage in Japan.

As many people are now online, they now have access to more enhanced social capital (Livingstone and Brake 2010; Adkins 2009; Perry-MacLean 2010). Social networks that are found on social media are now a significant component of social capital. The personal contacts and links to other people can help the members of a social network build social capital by offering each other access to informational resources, social status and power (Burgess 2009). Social media users have social assets through their community, which is created and maintained by themselves to develop mutually beneficial relationships.

As social support is another important element of social capital, previous research has indicated how it can be observed to happen often online, as online support does not involve physical or necessarily require monetary transactions (Livingstone and Brake 2010; Perriton 2008). Such support seen online could be in the form of information, emotional support and self-help support (Notley 2009; boyd and Ellision 2007).

The final component of social capital in social networks is trust. As new social media can provide better accountability of the user, one can anticipate more honesty among users and more cooperative behavior. The social networks in social media can potentially lead to more trust. Social media can be used to increase formal and informal networks as links can become stronger and therefore provide space for deliberation and social connections.

Regarding the communication activities of Twitter, the two separate types of bridging and bonding through social capital may both be present. The communication of bonding among close individuals may occur in Twitter, but the bridging of diverse individuals is also equally possible through Twitter, so the intermix of the two types of building social capital should also be considered. In the subsequent hours and days of the Great East Japan Earthquake, the wider usage and adoption of electronic networks such as social media were observed in areas of mid-level disruptions throughout Japan. Social media users had greater interaction among other online individuals and the growth of a social support communication network was observed. Information about lifelines and support, that is always in great demand by disaster victims had been very difficult to transmit in previous earthquakes. In the Great East Japan Earthquake, social media may have been able to provide and supply information and knowledge to socially support many disaster victims.

Social Support: Information Through Twitter During Emergency Situations

Previous studies have indicated how social media can be a backchannel to communicate information that cannot be received through the traditional media (McCarthy and boyd 2005; Sutton et al. 2008). Twitter has been discussed as a possible platform for providing and sharing information during emergencies (Hughes and Palen 2010; Mills et al. 2009). White (2010) has suggested how Twitter can effectively communicate the severity and range of a disaster, by linking documents and pictures and also being able to transmit location if using a mobile device. Information is posted almost real time, so logistic information can be retransmitted to rescue teams during crises.

During the Great East Japan Earthquake, traditional media and websites could not provide information about lifeline disruptions or other necessary information for the vast majority of victims in disrupted areas. The whole prefecture of Ibaraki had severe damage; however, this was almost completely ignored because Ibaraki prefecture does not have a local commercial television station belonging to a national network, and NHK (the public broadcaster in Japan) in Mito was not able to relay information of the destruction in the prefecture, as the area surrounding and including Mito experienced widespread destruction and disruption by the earthquake. The widespread destruction in Iwate, Miyagi and Fukushima had eclipsed most of the damage in Ibaraki. Therefore, the traditional media outlets of television, radio and newspapers were highly ineffective in supporting the people living in Ibaraki prefecture in the early days of the disaster.

So how could people get information about disaster relief in Ibaraki? For example, in the city of Tsukuba, vital information was provided through the city of Tsukuba's Twitter account, as detailed in my recount of events earlier. The 2011 White Paper of Information and Communications in Japan (Ministry of Internal Affairs and Communications, Japan 2011) concurs that Twitter played a major role in disseminating information during the disaster in many Japanese communities, as the white paper shows how the number of followers and tweets per day of 11 local government Twitter accounts affected by the disaster and the 28 peripheral local government accounts each increased tenfold after March 11, 2011. Many Twitter users also began following newspapers and local radio and television stations in the disaster areas that had Twitter accounts.

In this section, we analyze the role of Twitter during natural disasters and examine the dynamics of social capital building through the interactions by social media usage, focusing on the Tsukuba municipality. While the damage in Ibaraki Prefecture was relatively small in comparison to Miyagi, Iwate and Fukushima Prefectures, electricity, water supplies and transportation were all severely disrupted for an extended period of time in Tsukuba. During the period immediately after the initial earthquake, from March 11 until March 17, the City of Tsukuba Information Systems Department continued to transmit information through their Twitter account, @tsukubais, until normal computer systems and network connectivity were restored. In order to analyze the role of Twitter during this disaster in this area, I conducted a content analysis of all the tweets in the seven days after the March 11, 2011 earthquake until March 17, 2011 to view the trends of tweets and retweets. I also conducted an exploratory interview on April 25, 2011 with the account holder of the Twitter account, @tsukubais.

Figure 2.1 indicates the frequency of tweets and retweets by the Tsukuba Information Systems Department (currently renamed the Information Policy Department) Twitter account from March 1 to March 17, 2011. The preceding ten days were included for contrast.

As one can observe, the number of tweets increase sharply after the initial earthquake and the frequency is continuously higher than the daily number of tweets preceding the earthquake. The total number of tweets and retweets after the earthquake on March 11, 2011 until restoration of the computers of Tsukuba City Information Systems on March 17, 2011 was 593. Through analysis by text analytics computer software, all terms that construct meaning were extracted from the tweets and categories were built to measure frequency of common concepts in the tweets. Table 2.1 indicates the frequency of terms that were relevant to the disaster and measures taken. (I have translated the Japanese terms into English equivalents.) Table 2.2 indicates the frequency of terms that are relevant to the function of the Twitter platform.

One can observe that terms related to transportation, water and electricity supplies were used frequently; however, so were the terms related to calling out to people, expressions of gratitude and cooperation. The text analysis results provide evidence that the Tsukuba City Twitter account holder was having interactivity and communication with its followers. The relatively high frequency of tweets from specific smartphone software (Echofon Twitter software) indicates how Twitter was being used from a smartphone in this period.

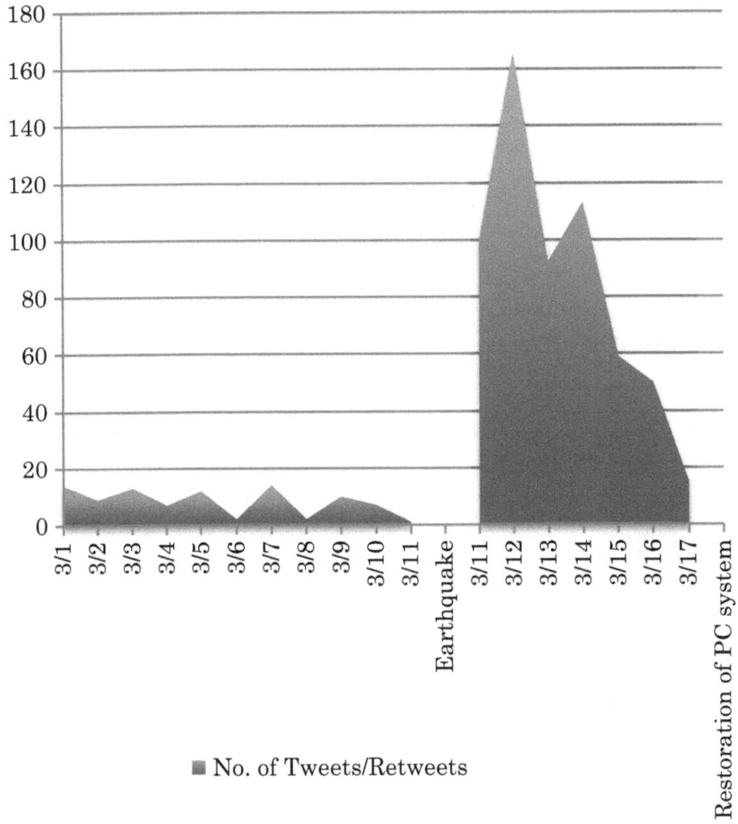

Fig. 2.1 Number of tweets and res (March 1–17, 2011)

Table 2.1 Frequency of terms relevant to the disaster and measures taken

High-frequency terms in tweets in relation to the disaster and measures	*Count*
Transportation (i.e., tx, schedule, operation, bus)	137
Calling out to people (i.e., please help, everyone)	118
Terms indicating time (i.e., now, presently)	102
Information on water (i.e., water supply)	77
Information on electricity (i.e., outages)	46
Expression of gratitude (i.e., thank you)	44
Expression of cooperation	25
Tweets by a Tsukuba city council member	15

Table 2.2 Frequency of terms that are relevant to the function of the Twitter platform

High-frequency terms in tweets in relation to the functions of Twitter	*Count*
rt, Retweet (in response to Tweets)	247
Smartphone software usage (i.e., echofon)	94
Links to other websites	72

Interview with Tsukuba City Twitter Manager

For further examination and clarification of the actual interactions from this Twitter account, on April 25, 2011, I conducted an exploratory interview with the Tsukuba City Information System Department Twitter account manager at the time of the earthquake, Mr. S. This was approximately six weeks after the March 11, 2011 earthquake in Tsukuba City. The following is a summary of the content of the interview.

The Tsukuba City Information Systems Department (currently renamed the Tsukuba City Information Policy Department) had been conducting an experiment to examine the effectiveness of using Twitter to communicate with its citizens from the latter half of 2010 through the account @tsukubais. In result, with the ongoing experimental period prior to the March 11, 2011 earthquake, @tsukubais already had approximately 2000 followers just before the initial giant earthquake hit Northeastern Japan. Immediately after the earthquake, Tsukuba City created a disaster countermeasures headquarters in the city hall. The damage to the computer and communications infrastructure in the city hall building was extensive, meaning that it was impossible to update the Tsukuba City website for several days. The Tsukuba City Information Systems Department decided to transmit any information that was being aggregated at the disaster countermeasures headquarters to the citizens via Twitter and also started collecting information from citizens through Twitter. Several members of the Tsukuba City Information Systems Department were on a business trip outside of Ibaraki, so Twitter was used to communicate with these members on how to get back to Tsukuba. The account manager also started communicating information to citizens who were unable to return to Tsukuba due to the wide disruption of public transportation systems throughout eastern Japan.

Information and updates on the electricity and water outage situation was being transmitted continuously, along with information about public transportation and the major roadways in and out of Tsukuba city. The water outage was predicted to take a longer time to repair in comparison to the recovery of electricity power facilities, therefore information and know-how in relation to emergency drinking water and toilet facilities from disaster victims of previous giant earthquakes began to be posted on the @tsukubais account. Such information was being relayed out through the Twitter account to the followers and readers.

What was noteworthy, and something of an anomaly, was the speed of information coming out of the municipality immediately after the disaster. Any useful information coming into the disaster countermeasures headquarters was being quickly distributed out to the citizens through Twitter from the start of the disaster. The department head of the time was taking full responsibility of bypassing normal procedures, such as those that required the sending of circulars around for everyone's stamp of approval in the government before tweeting anything. In result, emergency related information was being communicated at an unprecedented speed for a Japanese municipality. As aftershocks persisted throughout the following week after March 11, 2011, anxiety levels of government workers and citizens were high, so the account manager took extreme care on using language expressions not to further heighten anxiety. The 140-character upper limit required the account manager to carefully plan how to transmit information in easy-to-understand, normal language, without using any technical terms. Other information about Tsukuba was being tweeted through other accounts, so the #tsukuba hashtag was being used to facilitate those following topics in this context. The use of normal language enhanced the feeling of affinity towards to the account manager of the Tsukuba municipality among many of the followers, so an interactive environment for communication with the citizens using Twitter was created in the days following the earthquake. Twitter also allowed person-to-person communication or one-to-many communication depending on the situation. When the account manager needed to answer the same question that was being asked by many people (i.e., questions on the predicted time of recovery of the water supply, etc.), Twitter allowed for efficient and effective communication.

After three days, the account manager had other workers in the Tsukuba Information Systems Department help him tweet to the citizens, as the

Twitter account became a main channel of information until the Tsukuba City main website could be restored. The number of followers increased from 2000 to over 10,000 in a few days following the first big earthquake on March 11, 2011, and many followers were able to contribute information and know-how on methods of coping with this type of large-scale disaster. For example, knowledge on building makeshift toilet facilities or other techniques that can be used during water shortages were shared through Twitter among the followers of the account. A member of the city council created a makeshift volunteer group to translate the tweets coming out of this account into English, Chinese and Korean to the comparatively large foreign population living in Tsukuba due to its university and many national research institutes being clustered in Tsukuba. Cooperation among citizens, promotion of volunteer activities and other information sharing was being made available through Twitter to help the disaster countermeasures efforts of the municipality. Many messages of gratitude and communication helped boost the morale of the Information Systems Department and City Disaster Headquarters of the municipality. In reflection, the account manager also pointed out how it was important to always include a date and time into the tweet, just to make sure that people wouldn't confuse older tweets as being real-time information, such as information of the hours of operation of emergency supply vehicles for distributing water.

Twitter and Rumors During Disasters

Although the pro-social effects of Twitter use were verified in the Tsukuba city case during the Great East Japan Earthquake, social media contributed to the dissemination of false rumors as well during the same period. In the case of Tsukuba, the Twitter account manager worked hard to negate any false rumors being circulated on Twitter during the crisis, however, because information is being communicated so rapidly among Twitter users, false rumors unfortunately spread quite rapidly as well. Subsequently, this information becomes widely spread throughout by email, but at a much slower rate. One noteworthy and highlighted rumor that spread immediately after the Great East Japan Earthquake dealt with the fire that broke out in the industrial complex of Cosmo Oil Co Ltd. The essential content of the rumor was disseminating false information claiming that hazardous materials would begin to fall from the sky with the rain, and that the information source of this imminent danger was from employees working at the complex. Tweets and retweets of this information quickly spread during the hours right after the earthquake. On March 12, 2011 the disaster headquarters officially negated such rumors by tweeting information coming from the Chiba prefectural government (Chiba Prefectural Government 2011).

In the 2011 White Paper of Information and Communications in Japan (Ministry of Internal Affairs and Communications 2011), a keyword analysis using Google Real-time Search was conducted for investigating the false rumors during the Great East Japan Earthquake. The study averaged out the count or number of emergence of a keyword per minute, throughout the duration of the false rumors of Cosmo Oil. The keyword "Cosmo Oil" begins appearing in tweets right after the initial earthquake on March 11, 2011 and continues increasing from evening until midnight, at which the number of tweets reaches a peak and then begins to fall. From around 11 a.m. on March 12, 2011, the number of tweets begins to increase again. However, from around 3 p.m., the tweets that are negating the rumor begin to increase, and by March 13, the false rumors decrease dramatically. From this analysis result, one can observe that the spread of false rumors was very rapid through Twitter, but once tweets that negate the false rumor begin to appear, the false rumors also decrease very quickly on Twitter. Although the rumors no longer were circulated through Twitter, these false rumors subsequently continued to slowly spread by email, and this took a lot longer to negate in comparison to what happened on Twitter.

Although the Cosmo Oil rumor was a false rumor that began to spread in the wake of the disaster when people were still in panic, even after several months, many kinds of rumors, such as those related to radiation contamination continued to appear. Right after the March 11, 2011 earthquake, drinking water, foods with long shelf life, gas cartridges, gasoline fuel, toilet paper and tissue paper were in shortage because people began receiving false information that these supplies were running short due to disruption of logistics. This caused panic among people in many areas, and people started hoarding these items until it eventually created a chain reaction. Such chain reactions have happened in the past in Japan (i.e., the oil shock in the 1970s) and were deemed ridiculous, however the reporting of this behavior on mass media and Twitter simply aggravated the problem right after the earthquake. This unnecessary hazard occurred due to human behavior and incorrect information and irresponsible communication, sending a negative sentiment to the population that these supplies would be in shortage due to the earthquake. As the rumor grew, many assumed a possible problem would truly happen with commercial logistics in the weeks to come. Twitter account holders would tweet or retweet this type of information without verifying the truth of the information being communicated. The account holders did not perceive the grave consequences of transmitting this type of information, so no malicious intent was possibly perceived in them, however the result was a breakdown in these supplies in eastern Japan during the month after the earthquake. The breakdown even affected relief activities in the areas of dire need. In other instances, some tweets had malicious intent, such as those that swindled relief donations for the disaster victims. Through these incidents, we clearly know that Twitter in some instances, like these that were observed after the Great East Japan Earthquake, may cause unfavorable consequences for society.

Social Media and Information of the Fukushima Daiichi Nuclear Accident

The Great East Japan Earthquake of March 11, 2011 caused various types of enormous damage throughout Japan, but the severely damaged Fukushima Daiichi Nuclear Power Plant is the most catastrophic and has left a terrifying after-effect that continues to this day. The tsunami waves caused by the earthquake managed to destroy the diesel backup power systems at the Fukushima Daiichi nuclear power plant, which led to the sequential meltdown and explosions of the reactors in the days following the earthquake. The residents living in the vicinity of the Fukushima Daiichi Nuclear Power Plant had to evacuate during and after the accident in March 2011. Several systems were implemented by the governmental agencies to assist effective evacuation of residents in these situations. One of these systems is named the Development of the System for Prediction of Environmental Emergency Dose Information (SPEEDI). Acting as an academic advisor during the preparation of the Nuclear Risk Research Center of Tokyo in from 2014 to 2015, I reconfirmed and verified much of the following information in regards to SPEEDI that occurred after the Fukushima Daiichi Nuclear Power Plant accident that has also been heavily covered by the Japanese mainstream media in the past.

SPEEDI was initiated after the nuclear power plant accident that occurred at the United States Three-Mile Island nuclear reactor in

1979. The actual development of the system begun in 1980 by the Japan Atomic Energy Research Institute and has cost over 10 billion Yen over the course of its improvement and development. The base system of this prediction system was completed by 1984 and development has continued on improving the system since the initiation of the project. The system had been continuously updated in accordance to the latest technical, scientific and technological developments and the current system has been in operation since 2005 by the Nuclear Safety Technology Center. The Japanese Ministry of Education, Culture, Sports, Science and Technology (MEXT) had been responsible for overall administration of SPEEDI, so that it can be quickly activated to predict the approximate range and density of radioactive contamination in the air and approximately calculate the level of radioactive dosage when in a potential emergency situation or during an actual accident of a nuclear facility that is releasing radioactive particles into the environment. SPEEDI can output results according to the available input of information about the radiation being emitted by the radiological source event and include weather and geographical patterns into its calculations to create a fairly accurate image of the radioactive fallout and contamination.

The Act on Special Measures Concerning Nuclear Emergency Preparedness of Japan indicates the procedures in the basic plan for disaster countermeasures and how environmental radiation monitoring needs to be conducted in accordance to law in the case of a radiological accident. According to MEXT, SPEEDI can produce graphic map-like figures depicting its calculations and predict the range of radiological particles. The diagram to illustrate the calculation results is designed so that it is simple to interpret and be understood by non-experts of radiological events. These predictions can help governmental policy making in determining the scale of evacuation and make the best decision on possible escape routes for residents to lower any extra accidental dosage of radiation in the case of an actual disaster situation. The calculation results can be also used for deciding how and when to distribute iodide tablets to prevent the Iodine-131 thyroid gland accumulation health risk in such an event. MEXT had devised a disaster countermeasure operations plan that switches normal SPEEDI operations into emergency mode, accompanying the radioactive emergency alert in accordance to Article 10 (*Jyu-jyo tsuho*) of the Act on Special Measures Concerning Nuclear Emergency Preparedness of Japan established in 1999.

According to the Ministry of Justice (2012) of Japan, Article 10 states that:

> When a nuclear emergency preparedness manager has been notified that a radiation dose above the limit specified by a Cabinet Order has been detected, pursuant to the provisions of a Cabinet Order, near the border of an area where the nuclear site is located or has discovered such fact for him/herself, he/she shall, pursuant to the provisions of an ordinance of the competent ministry and the nuclear operator emergency action plan, immediately notify the competent minister, the competent prefectural governor, the competent mayor of a municipality and the related neighboring prefectural governors (in the case of the occurrence of an event pertaining to transport outside the nuclear site, the competent minister, and a prefectural governor and the mayor of a municipality who have jurisdiction over the place where said event has occurred) to that effect. In this case, the competent prefectural governor and the related neighboring prefectural governors shall notify the mayors of related surrounding municipalities to that effect.

During emergency mode, SPEEDI begins calculations predicting the spread of radiation and the results are shared through devices owned by the Nuclear Safety Commission, Nuclear Industrial Safety Agency, MEXT, prefectural governments and other municipalities with nuclear facilities, off-site centers (OFC, which become emergency countermeasure centers in the case of an accident) of nuclear facilities. The procedure for SPEEDI to enter into emergency mode requires approximately 30 minutes for data input to have results being produced for predicting the following 24 hours after initiating the mode. If the information about the amount of release of radiological particles from the contamination source is available, calculations can be made using that information to augment results. Otherwise, the predictions are based on a calculation hypothesizing that 1 Bq (Becquerel) or any assigned amount per hour is being released from the radiological source.

The Act on Special Measures Concerning Nuclear Emergency Preparedness of Japan clearly indicates that in the event of an emergency, these results are to be used in supplying the government and municipalities of detailed information about the possible radiological fallout from an accident. This information was to be shared by all citizens of Japan, help everyone prevent unnecessary exposure to radiation and protect the health and lives of younger children susceptible to the effects of Iodine-131 to the thyroid gland, which has been documented during the Chernobyl Nuclear Power Plant accident.

During the immediate hours and days that followed the Great East Japan Earthquake and the Fukushima Daiichi nuclear power plant accident, the government did not release the calculations of SPEEDI to the public, even though the system was set into emergency mode two hours after the first large earthquake that occurred at 14:46 JST. The Emergency Response Support System (ERSS) that was supposed to send exact data from the power plant was not responding, therefore hypothetical data was being used for SPEEDI to predict possible patterns of contamination. The system worked successfully and produced thousands of possible patterns that correctly illustrated the direction in which radioactive particles would spread. Following the results of potential widespread contamination being calculated by SPEEDI, MEXT initiated actual radiation monitoring of areas predicted to have high radiation measurements by their monitoring cars. On March 15, 2011, the measurements by these monitoring cars indicated 300 microSieverts per hour in Namiemachi of Fukushima Prefecture. MEXT did not immediately disclose the precise name of Namiemachi, which location was producing measurements with these high levels of radioactive doses, and avoided any clear indication of this spot on the maps in their announcements. What happened in result, was having MEXT release measurement information of an undisclosed area with very high doses radiation and subsequently, did not warn the residents of Namiemachi of this situation. As this specific location was outside the initial 20 km perimeter for evacuation, some residents ended up remaining there until April 11, 2011. The Nuclear Industrial and Safety Agency (NISA) also had acquired the results of SPEEDI, but they were unable to use these results effectively as well. As a result, the Japanese cabinet under Prime Minister Naoto Kan and the government did not plan evacuations based on the calculation results by SPEEDI and instead, set a circular evacuation perimeter surrounding the Fukushima Daiichi Nuclear Power Plant.

The topic of SPEEDI began appearing in the Internet forums, Twitter tweets and Facebook and other SNS discussions following the March 23, 2011 mainstream media news reports (a delay of more than ten days) made by the media organizations. There is no evidence of wide spread discussions about SPEEDI information leaked prior to the news organizations on a large scale. However, the information on radioactive fallout being provided by non-Japanese governmental organizations outside of Japan were being linked or quickly translated and uploaded onto the forums and other social media such as Twitter and Facebook.

The United States Nuclear Regulatory Commission and United Kingdom Scientific Advice Group in Emergencies were making assessments of the situation very early, and were quick in making the information public via websites and other information outlets, so this information was being translated into Japanese and relayed to the Japanese public via social media such as Facebook, Twitter and YouTube other means such as websites through the Internet. At the same time, L'Institut de Radioprotection et de Sûreté Nucléaire of France (IRSN) was also announcing accumulation of radiation by Fukushima on their website. Zentralanstalt für Meteorologie und Geodynamik (ZAMG) of Austria that also had been conducting surveillance of the Comprehensive Nuclear Test Ban Treaty was predicting the spread of Iodine-131 from Fukushima according to the weather patterns and atmospheric movement from March 15, 2011. The information of IRSN and ZAMG was also being relayed among forums and various social media in Japan. The Deutscher Wetterdienst and Norwegian Institute for Air Research Department of Atmospheric and Climate Research were having their early reports transported onto YouTube and Japan's Internet video sharing platform Nico Nico Douga. These were also being accompanied by news reports on how German citizens living in Tokyo and Yokohama were being advised to evacuate, raising concern about the conflicting reports being announced by the Japanese government at the time concerning radiological contamination. Japanese social media such as Twitter and SNS users of Facebook were also featuring the consultant Kenichi Ohmae's live interviews being relayed on YouTube concerning his views on the Fukushima reactor, early predictions of a reactor meltdown, radioactive exposure dosage and policy failure of the Japanese government.

Through these various sources, along with the information being provided by the United States Armed Forces stationed in Japan, information was quickly being relayed through social media such as Twitter and Facebook, translated into Japanese, and then being received by a growing audience in Japan. Although the information of SPEEDI was not being conveyed as quickly as anticipated in its original scheme and design, outside governmental sources were providing Japanese social media users sufficient information for predicting the possible patterns of contamination of radiation during the first weeks of the Fukushima Daiichi nuclear power plant accident, prior to actual the release of SPEEDI predictions.

One noteworthy discrepancy that discredited the Japanese government's competency was the evacuation perimeter designated by the United States Nuclear Regulatory Commission and how the Japanese Government and media reacted in the days that followed March 11, 2011.

US citizens in a 50 mile (80 km) radius were advised to evacuate based on the studies based on Chernobyl, while the Japanese government was advising an expansion of the perimeter from 3 kilometers to 10 kilometers, to 20 kilometers on March 12, 2011. United States Armed Forces radio announcements in the days following the explosions at the Fukushima Daiichi Nuclear Power Plant were advising non-active personnel in the Kanto area to try to stay indoors and minimize risk to radiation exposure as a precaution while the Japanese government did not make any similar types of announcements. Ten months after the accident, MEXT Minister Hirofumi Hirano announced on January 17, 2012 that the SPEEDI calculation results were sent to the United States Armed Forces in Japan through the Ministry of Foreign Affairs on March 14, 2011, which is nine days before it was made public to the Japanese.

Local residents and governments in Fukushima that had to evacuate the immediate vicinity did not have access to the calculation results of SPEEDI, and unknowingly led its evacuation of the residents into the northwest direction, which was subject to the highest level of radioactive contamination following the explosions and vents made by the Fukushima Daiichi nuclear power plant. If the calculation results of SPEEDI were readily available, local municipalities would have been able to avoid having their residents being exposed to unnecessary high levels of radiation in the following weeks due to the Fukushima Daiichi nuclear power plant accident. The reason why SPEEDI was not properly utilized may be due to fear of panic and unnecessary confusion that caused the policy makers in Japan to hesitate in disclosing the calculation results. Hiding the calculation results of SPEEDI was a big error in the process of crisis communication and offset the whole strategy of the Japanese government's attempt to contain the situation through half-baked tactics. This error, in retrospect was a human error that led to another set of future problems, as strong anxiety of not being able to receive accurate information on radiation contamination persists in Japan and in Fukushima.

During the time period where many areas in northeastern Japan were still left in devastation or being evacuated due to the Fukushima accident, many areas within Japan and outside began to panic about the risks of radiation. As substantial amounts of radioactive material were released during the accident at Fukushima Daiichi, many people became concerned of the health issues involved, such as acute health effects like the increase of thyroid cancer and internal radiation contamination. In retrospect, the health effects still remain controversial, but the evacuation and the sudden breakdown of communities have left social and psychological scars among the large number of people who experienced psychosocial distress by this accident.

In the days following the Fukushima accident, one noteworthy phenomenon in social media that occurred in Japan in relation to radiological contamination was the element of distance from Fukushima. The weather patterns after the Fukushima Daiichi nuclear power plant accident carried radiological material to a wide region of Eastern Japan and as information about the Fukushima Daiichi nuclear power plant meltdown and radioactive contamination was initially hidden, this led to panic about health safety, as most Japanese had very limited information or knowledge about radioactivity at that time. These concerns towards health safety, and food safety heightened in Tokyo and the surrounding suburbs, temporarily creating shortages of bottled water and other foods that were deemed to be uncontaminated. Taro Yamamoto, who is now a member of the Japanese Upper House, tweeted his concerns about children and radioactive contamination and his words were heavily circulated through retweets and through other social media. The noteworthy phenomenon here is about the location of users and the frequency of the social media use to communicate concerns about radiological contamination. Rather than the people who were in the immediate vicinity of the Fukushima Daiichi nuclear power plant and at the highest risk, those who were further away, in highly populated urban areas such as Tokyo and the suburbs that surround Tokyo, expressed grave concern about radiological contamination and voiced their opinions on social media. This somewhat peculiar phenomenon requires more research; however during the case of the Fukushima Daiichi nuclear power plant accident, grave concern about radiological contamination among the people in the areas that were at a distance of approximately 300 kilometers or more with trace radiological contamination was higher than those that were closer at 200 kilometers or less, and Japanese who were voicing their concerns were also more salient as well. When the distance became larger at 400 kilometers or more, concerns about contamination were observed to be lower. One can suggest population and stress to be other elements that could have affected the emergence of this phenomenon; however, this may be due to the areas that were 200 kilometers or less also experienced mid to high-level disruptions, destruction and damage and had more urgent things to worry about other than radiation at the time period right after the initial earthquake.

Crisis Communication, Social Media and Trust in Government and Mass Media

During the Great East Japan Earthquake, voice communication through mobile telephone was reconfirmed to be difficult in the affected areas; however, Internet access through mobile devices was found to be relatively robust and resilient in comparison to normal telecommunication channels. VoIP, SMS and email are quicker solutions for communication and overcoming telecommunication breakdowns in such disaster situations. Radio is more effective because of its relatively portability in comparison to televisions and also it can be battery-operated. Social media such as Facebook and Twitter was more effective in disseminating information on the Internet than websites early in a disaster. Twitter was found to be most effective for quickly spreading information whether correct or false as this chapter also examined how Twitter has the potential to transmit false rumors very rapidly during disasters. This requires some further consideration and one needs to devise how to avoid some of the unfavorable consequences that were observed during the Great East Japan Earthquake. False rumors were quickly negated on Twitter, although we also observed email takes longer for negation of false rumors.

The main portion of the first section of the chapter provides evidence that Twitter is a communication means that has characteristics that are greatly beneficial during disasters. Through the quantitative content analysis of "tweets" of the Tsukuba City Information Systems Department and interview with the account manager, Twitter was found to be raising social capital to new levels and it played a significantly pro-social role during the disaster. In the case of Tsukuba, many of the lifelines, such as water and electricity, were not functioning, but social media such as Twitter functioned for building communication lines that were trustworthy. A communication backchannel until the disaster became the main communication channel for the citizens of Tsukuba and the disaster countermeasures headquarters of the city.

Through social media, a communication network of social support and trust among the government and stakeholders was formed, and social interaction was made possible in a wide region. Twitter became the supplier of information and knowledge for the citizens of Tsukuba in the early days of the disaster and the basis for further support because social capital was high to begin with among many Japanese communities. As the earthquakes and havoc by the tsunami continued to plague Japan, looting was comparatively limited and no rioting was observed after the large earthquake. The Japanese social norms of maintaining order prevailed with a high level of social capital already present in Japan. The Japanese collectively found the necessity to maintain social order in this grave situation, and this was reflected onto the social media usage of the Japanese during this large-scale disaster in various regions. People retweeted useful information and communicated with each other. People who were in need of help, or who were in various predicaments, were helped by others through useful information being communicated and retweeted through the function of Twitter. Although the pre-existing high level of social capital in Japan was an advantage, this does not imply that the findings in this section are only limited to Japan. The means of transmitting and receiving information and building of trust observed in the first section of this chapter are effective, and can be applied to other disasters, in other nations and societies.

Initially, the experiment of using Twitter for communicating to the citizen of Tsukuba was received with little enthusiasm and mostly negativity among the heads of Tsukuba City Hall. In retrospect, thanks largely to the fact that the Information Systems Department of Tsukuba City stepped out of its bureaucratic bounds and embarked on sending out timely information, the tweets of the account manager became a major source of providing vital information on lifelines and infrastructure until restorations were made after the disaster.

On the other hand, in the second half of this chapter, I re-examined the debacle of information mishandling by the Japanese governmental agencies during the Fukushima Daiichi Nuclear Power Plant Accident that led to the general distrust of the Japanese government. The nuclear accident overwhelmed Tokyo Electric Power Company (TEPCO), many of the agencies and ministries. It exposed the lack of preparedness towards nuclear accidents. In retrospect, the delay in providing information of radiological contamination that was already in hand through SPEEDI led to the radioactive exposure to many residents who evacuated in the direction that was affected by the Daiichi Nuclear Power Plant Accident. To this day, the credibility of Tokyo Electric Power Company is very low since the accident, and distrust is evident in information regarding radiation provided by the government and regulatory agencies. The mismanagement of the crisis and communication strategy of hiding information has forever damaged the reputation of TEPCO. Social activism that was unseen in

Japan for a long period of time emerged again and nuclear energy policy was the new adversary. As the initial information release by the Japanese government regarding the meltdown was so slow, not only the citizens of Japan but other governments as well made preemptive evacuation plans for their citizens residing in Japan immediately after the nuclear accident. The strategically created image of nuclear energy as a clean energy source has been wiped out. Japanese activists continue to protest the restart of Japanese nuclear power plants that were suspended after the Fukushima accident.

Many Japanese people that searched for information through social media and the Internet for information on radiological contamination no longer fully trust announcements from the Japanese government or the mass media. Although they may use the information found on mass media as a reference, they will try to find other information sources and discuss this on Twitter or other social media. Sole reliance on mass media in Japan changed from March 11, 2011. Standard strategic communication plans needed to be changed. People experienced the need to look and search for information that was not available through the traditional media, and social media was the quickest to supply information to this new demand. People can acquire information through social media, think and decide on what to do. The Japanese who until that day, who were avoiding the murky, dangerous and sinister Internet could no longer afford staying offline. The Japanese people were thrust into an environment where one was required to search information for their personal safety and well-being of their family. Social media was the solution. YouTube provided information of the disaster. Nico Nico Douga was transmitting and relaying television broadcasts to those without television reception in March 2011 after the initial earthquake and also transmitted information reports on the damage caused by the earthquake from various regions in Northeastern Japan. Residents in the devastated areas were able to send out information through social media, and public offices and media began using it as well. Inquiries on the safety of people were sent through social media. Direct messaging among users of Twitter became a standard means of communication. Information about volunteering was also shared through social media. The Great East Japan Earthquake and Fukushima Daiichi Nuclear Power Plant accident became a major catalyst in shifting many Japanese to begin using social media and pushed away previous concerns towards using the Internet (see Chap. 1).

References

Adkins, A. M. (2009). *Myspace, Facebook and the strength of Internet ties: Online social networking and bridging social capital*. Unpublished master's thesis, University of Akron.

Bargh, J. A., & McKenna, K. Y. A. (2004). The Internet and social life. *Annual Review Psychology, 55*, 573–590.

Baym, N. (1995). The emergence of community in computer-mediated communication. In S. G. Jones (Ed.), *CyberSociety: Computer-mediated communication and community* (pp. 138–163). Thousand Oaks: Sage.

Bourdieu, P. (1986). The forms of capital. In J. G. Richardson (Ed.), *Handbook of theory and research for the sociology of education* (pp. 241–258). New York: Greenwood Press.

boyd, D. M., & Ellison, N. B. (2007). Social network sites: Definition, history and scholarship. *Journal of Computer-Mediated Communication, 13*(1), 210–230.

Burgess, K. R. (2009). Social networking technologies as vehicles of support for women in learning communities. *New Directions for Adult and Continuing Education, 2009*, 63–71.

Chen, W., Boase, J., & Wellman, B. (2002). The global villagers: Comparing Internet users and uses around the world. In B. Wellman & C. Haythornthwaite (Eds.), *The Internet in everyday life* (pp. 74–113). Oxford: Blackwell.

Chiba Kencho [Chiba Prefectural Government]. (2011). *Kosumo Sekiyu (Kabushikikaisha) no tank kasai niyoru fuuhyou nituite* [Concerning the false rumors of the Cosmo Oil tank fire]. Retrieved http://www.pref.chiba.lg.jp/shoubou/20110312-fuuhyou.html

Cohen, D., & Prusak, L. (2001). *In good company: How social capital makes organizations work*. Boston: Harvard Business School Press.

Ellison, N. B., Steinfield, C., & Lampe, C. (2007). The benefits of Facebook "friends:" Social capital and college students use of online social network sites. *Journal of Computer-Mediated Communication, 12*(4), 1143–1168.

Ellison, N. B., Steinfeld, C., & Lampe, C. (2011). Connection strategies: Social capital implications of Facebook-enabled communication practices. *New Media & Society, 13*(6), 873–892.

Granovetter, M. (1973). The strength of weak ties. *American Journal of Sociology, 78*(6), 1360–1380.

Helliwell, J., & Putnam, R. D. (2004). *The social context of well-being, Series B, 359* (pp. 1435–1446). London: Philosophical Transactions of the Royal Society.

Howard, M. M., & Gilbert, L. (2008). A cross-national comparison of the internal effects of participation in voluntary organizations. *Political Studies, 56*(1), 12–32.

Hughes, A. L., & Palen, L. (2010). Twitter adoption and use in mass convergence and emergency events. *International Journal of Emergency Management, 6*(3–4), 248–260.

Joinson, A. N. (2008, April). *Looking at, looking up or keeping up with people? Motives and use of Facebook*. Paper presented at the 26th annual SIGCHI conference on Human Factors in Computing Systems, Florence, Italy.

Kobayashi, T., & Ikeda, K. (2005). Onrain komyunitii no Shakaikanekishihon [Social capital in online commnities]. In K. Ikeda (Ed.), *Intanetto komyunitii to Nichijosekai [Internet communities and daily life]* (pp. 148–184). Tokyo: Seishin Shobo.

Livingstone, S., & Brake, D. R. (2010). On the rapid rise of social networking sites: New findings and policy implications. *Children & Society, 24*(1), 75–83.

McCarthy, J. F., & boyd, D. M. (2005). Digital backchannels in shared physical spaces: Experiences at an academic conference. In *Proceedings of the SIGCHI Conference on Human Factors in Computing Systems (CHI '05) Extended Abstracts* (pp. 1641–1644). New York: ACM Press.

Mills, A., Chen, R., Lee, J., & Rao, H. R. (2009). Web 2.0 emergency applications: How useful can Twitter be in emergency response? *Journal of Information Privacy & Security, 5*(3), 3–26.

Ministry of Internal Affairs and Communications, Japan. (2011). Heisei 23 Nendo Johou Tsuushinn Hakusho [2011 White paper of information and communications in Japan]. Retrieved from http://www.soumu.go.jp/johotsusintokei/whitepaper/ja/h23/pdf/23honpen.pdf

Ministry of Justice Japanese Law Translation. (2012). Retrieved from http://www.japaneselawtranslation.go.jp/

Norris, P. (2001). *Digital divide: Civic engagement, information poverty and the Internet*. Cambridge: Cambridge University Press.

Notley, T. (2009). Young people, online networks, and social inclusion. *Journal of Computer-Mediated Communication, 14*(4), 1208–1227.

Perriton, L. (2008). *Women get wolves, men get mentors: Leadership development viewed from a social relations perspective*. Paper presented at the Engendering Leadership Conference, Perth, Australia.

Perry-MacLean, S. (2010). *Developing interprofessional collaboration using web-based instruction: A case study of the Healthy Minds/Healthy Children project*. Unpublished doctoral dissertation, University of Calgary (Canada), Canada. Retrieved from Dissertations & Theses: A&I. (Publication No. AAT NR64129).

Pfeil, U., Arjan, R., & Zaphiris, P. (2009). Age differences in online social networking—A study of user profiles and the social capital divide among teenagers and older users in MySpace. *Computers in Human Behavior, 25*(3), 643–654.

Putnam, R. (2000). *Bowling alone: The collapse and revival of American community*. New York: Simon & Schuster.

Quan-Haase, A., & Wellman, B. (2005). Hyperconnected net work: Computer mediated community in a high-tech organization. In C. Heckscher & P. Adler (Eds.), *Collaborative community in business and society*. New York: Oxford University Press.

Raacke, J., & Bonds-Raacke, J. (2008). MySpace and Facebook: Applying the uses and gratifications theory to exploring friend-networking sites. *Cyberpsychology & Behavior, 11*(2), 169–174.

Raine, L., Horrigan, J., & Cornfield, M. (2005). *The Internet and campaign 2004.* Washington, DC: Pew Internet and American Life project Report.

Scheufele, D. A., & Shah, D. V. (2000). Personality strength and social capital: The role of dispositional and informational variables in the production of civic participation. *Communication Research, 27,* 107–131.

Schuler, D. (1996). *New community networks: Wired for change.* New York: ACM Press.

Smith, A., & Raine, L. (2008, June 15). *The Internet and the 2008 election.* Washington, DC: Pew Internet and American Life Project. doi:http://www.pewinternet.org/pdfs/PIP_2008_election.pdf

Steinfield, C., Ellison, N. B., & Lampe, C. (2008). Social capital, self-esteem, and use of online social network sites: A longitudinal analysis. *Journal of Applied Developmental Psychology, 29*(6), 434–445.

Sutton, J., Palen, L., & Shklovski, I. (2008, May). Backchannels on the front lines: Emergent uses of social media in the 2007 Southern California Wildfires. In F. Fiedrich & B. Van De Walle (Eds.), *Proceedings of the 5th international ISCRAM conference* (pp. 624–632). Washington, DC.

Takahashi, T. (2010). Myspace or Mixi? Japanese engagement with SNS (Social Networking Sites) in the global age. *New Media and Society, 12*(3), 453–475.

Valenzuela, S., Park, N., & Kee, K. F. (2009). Is there social capital in a social network site?: Facebook use and college students' life satisfaction, trust, and participation. *Journal of Computer-Mediated Communication, 14*(4), 875–901.

Valkenburg, P. M., & Jochen, P. (2007). Online communication and adolescent well-being: Testing the stimulation versus the displacement hypothesis. *Journal of Computer-Mediated Communication, 12*(4), 1169–1182.

Wellman, B., & Gulia, M. (1999). Virtual communities as communities: Net Surfers don't ride alone. In M. Smith & P. Kollock (Eds.), *Communities in cyberspace* (pp. 163–190). Berkeley: Routledge.

White, C. (2010, December 14–15). Social media and meta-networks for crisis mapping: Collaboratively building spatial data for situation awareness in disaster response and recovery management. Paper presented at *2010 Specialist Meeting Spatio-Temporal Constraints on Social* Networks, University of California, Santa Barbara, Center for Spatial Studies. Retrieved from http://www.ncgia.ucsb.edu/projects/spatio-temporal/docs/White-position.pdf.

Williams, D. (2006). On and off the 'net: Scales for social capital in an online era. *Journal of Computer-Mediated Communication, 11*(2), 593–628.

Muneo Kaigo is an associate professor in the Faculty of Humanities and Social Sciences at the University of Tsukuba in Japan. He teaches courses in media communication in the Graduate School of Humanities and Social Sciences, media management in the Graduate School of Business Sciences and communication science in the College of Comparative Culture at the University of Tsukuba. He is currently leading a joint research project on social media uses among local municipalities in Japan with cooperation from the municipal government of Tsukuba and Intel Corporation Japan.

CHAPTER 3

Japanese Local Government Facebook Profiles

Sae Okura and Muneo Kaigo

Political Participation Through Social Media

The importance of the growth of the public sphere and cooperation between public and private sectors has been the subject of considerable debate. Cooperation is defined as a state in which "multiple actors inclusive of individuals and groups cooperate on an equal basis while complementing each other's different abilities and roles in order to achieve certain purposes" (Nishio 2004, p. iii), and this has been emphasized, in particular, in the context of participatory democracy and new public management (NPM) (Kodagiri 2014, pp. 11–13).

Participatory democracy focuses on the educational aspect of political participation and civic engagement, which can make citizens more democratic and act more publicly, and can eventually make political systems more stable (Kabashima 1988, p. 41). This theory focuses on the effectiveness of education through political participation, as occurs, for example, through elections, election campaigns, lobbying, advocacy activities, and so on

S. Okura (✉)
Faculty of Humanities, Law and Economics, Mie University,
Tsu, Mie, Japan

M. Kaigo
Faculty of Humanities and Social Sciences, University of Tsukuba,
Tsukuba, Ibaraki, Japan

M. Kaigo (ed.), *Social Media and Civil Society in Japan*,
DOI 10.1007/978-981-10-5095-4_3

(Pateman 1970; Kabashima 1988, pp. 39–44). Rochon and Kabashima (1998) found that African-American citizens, who finally obtained the right to vote after a long civil rights movement, use media to collect political information and information on internal political efficacy through political participation, indicating that political participation has positive effects on the education of citizens. Ikeda et al. (2008) also found that political participation can enhance the political efficacy of citizens. In other words, political participation through cooperation can create democratically minded citizens who are more aware of public issues.

One of the ways to integrate the government sector with the private sector is through the use of social media such as Facebook. Social media has the potential to create a useful environment for citizens so they can participate virtually and actively towards the promotion of a more open government (Lee and Kwak 2012). Citizens can become the producers of public services through social media (Linders 2012). Social media can also be used in times of crises (Kavanaugh et al. 2012); for example, civil society can assist the government during large-scale disasters that temporarily paralyze governmental functions. Opportunities created by enhanced online communication allow for the creation of greater social capital and a more effective civil society. For instance, McAtee and Wolak (2011) found that the existence of social networks leads to a higher degree of motivation and participation in civil society. Such networks create a space for the civic-minded members of society to engage in daily online communication and interconnectedness (Wellman and Haythornthwaite 2002). Some studies have indicated that virtual communities expect to facilitate the actions of civil society through the development of online social networks (Jennings and Zeitner 2003; Livingstone and Markham 2008; Shah et al. 2001, 2005).

Despite the importance of the growth of the public sphere and cooperation between the public and private sectors, we have little knowledge about how local governments use Facebook. What is the major focus of local governments in Japan, and what policy areas are the focus of local government Facebook pages that accumulate more followers and civic engagement?

To answer these inquiries, this chapter investigates how local governments in Japan use Facebook. Our aim is to examine specific areas of policy that are the areas of focus for each of the local government Facebook pages. This chapter will investigate all 425 Facebook community pages operated by local governments in Japan.

Political and Civic Participation, Civic Engagement and the Internet, and Social Networking Services (SNS)

Political participation has been defined as the "activity by private citizens designed to influence government decision-making" (Huntington and Nelson 1976, p. 3). According to a more specific definition of political participation, proposed by Samuel P. Huntington and Joan M. Nelson (1976), the following definitions are the core of political participation:

1. It is actual activity and does not include psychological preference.
2. It is an activity of the public, and does not include activities conducted by bureaucrats, politicians, and lobbyists as occupations.
3. It includes activity designated to influence the government, and it does not matter if the means to influence the government are legal or illegal.
4. It does not matter whether it actually can influence the government.
5. It includes not only voluntary and autonomous participation, but also participation mobilized by someone else.

There is a similar concept called "civic engagement" which refers to political participation by citizens. Zukin et al. (2006) define civic engagement as an "organized voluntary activity focusing on problem solving and helping others. It includes a wide range of work undertaken alone or in concert with others to effect change," inclusive of participation increasingly encouraged by the schools and facilitated by parents and community organizations (Zukin et al. 2006, p. 7). Takahashi and Sato (2013), by contrast, define civic engagement as an "activity by private citizens designed to influence government and society in order to solve the public issues in the community" (Takahashi and Sato 2013, p. 8). Based on this definition, civic engagement includes activity to influence not only the political actors in a narrow sense such as administration and government inclusive of parliament, but also social actors such as community organizations and non-profits. The definition also divides civic engagement into three different stages based on the level of engagement by private citizens: (1) civic engagement initiated by government; (2) cooperation; and (3) autonomy. Civic engagement initiated by the government has stronger levels of participation or initiative by the government, whereas autonomy has stronger and more intense participation or engagement by private citizens. Cooperation occurs between them, and the government and private citizens cooperate on an equal basis to achieve certain purposes (Takahashi and Sato 2013). Cooperation here is defined as being "multiple actors inclusive of individuals and groups [who] cooperate on an equal basis while complementing each other's different abilities and roles in order to achieve certain purposes" (Nishio 2004, p. iii). There could be multiple relationships between government and citizens regarding cooperation. Citizens may provide public services as partners of government, and they may lobby or even strongly influence the government to achieve specific purposes (Young 2000). The collaboration of actors such as civil society organizations (CSOs) participating in the provision of public services raises concerns both about the concept itself and also about the realities of collaboration. Having citizens, CSOs and the government in an equal relationship is a potential concern for the administrative body. In addition, there is also the concern whether or not residents and organizations will become "subcontractors" of government through collaboration.

As the purpose of our chapter is not to redefine what cooperation is, it may be more useful here to simply understand cooperation through the definition provided by Nishio (2004).

The level and form of political participation or civic engagement could vary according to the area of policy. For instance, Eto (1998) pointed out that more civic engagement with regard to social welfare should be encouraged because the more society develops and matures, the more the requirements for social welfare have been diversified and individualized. Fisher et al. (2015) also argue that more active civic engagement or political participation in environmental policy is essential to achieve better democracy. On the other hand, Zukin et al. (2006) indicate that minority and immigrant young people are less likely to express their political opinions by contacting officials and expressing opinions via media than majority young people.

One useful way to promote political participation or civic engagement, as we discussed in Chap. 2, is through the Internet and social media. Even for the Internet, the difference in devices to be used depends heavily on citizen preference. Kobayashi and Ikeda (2005) indicate the difference in impact on political and social participation to be found between mobile phone email usage and computer email usage. They find that mobile phone e-mail usage has a minimal impact on participation in activities and discussions regarding five political issues, whereas computer email usage has a more significant impact. At the same time, mobile phone email is found to have a positive correlation with informal group participation. These findings suggest that mobile phone usage happens more often in a private informal social context, which will not necessarily promote formal participation that is widely open to society, along with political participation and civic engagement (Kobayashi and Ikeda 2005).

Regarding the impact of social media such as Facebook on political participation or civic engagement, Boulianne (2015) produces 36 studies assessing the relationship between social media usage and participation in civic and political life. Her results indicates that (1) the metadata she collected demonstrates a positive relationship between social media usage and participation, that; (2) only half of the coefficients are statistically significant because studies using panel data are less likely to report positive and statistically significant coefficients between social media use and participation compared to cross-sectional surveys; and, finally, that (3) the metadata also suggests that social media usage has minimal impact on participation in election campaigns. Kushin and Yamamoto (2010) examine college students' online media usage for political purposes during the 2008 US national election. Their results indicate significantly positive relationships between attention to traditional Internet sources and political self-efficacy and situational political involvement. However, they also discover that attention to social media is not significantly related to political self-efficacy or involvement, and online expression is significantly related to situational political involvement but not to political self-efficacy (Kushin and Yamamoto 2010).

On the other hand, according to Gil de Zúñiga, Jung, and Valenzuela's findings using US national data, offline and online use of traditional media, political constructs (knowledge and efficacy), frequency and size of political discussion networks, and seeking information via social networking sites have a positive relationship and can be significant predictors of people's social capital and civic and political participation, both online and offline (Gil de Zúñiga et al. 2012). According to Xenos, Vromen, and Loader's results based on cross-national data among young people in Australia, the USA, and the UK, there is a significant and positive correlation between social media use and political engagement among young people across all three countries (Xenos et al. 2014). Vissers and Stolle (2014) also provides evidence on how social media can be useful for political participation and civic engagement. According to Vaccari et al. (2015), based on an original survey of a representative sample of Italians who discussed the 2013 election on Twitter, the more respondents acquire political information via social media and express themselves politically on these platforms, the more they are likely to contact politicians via email, campaign for parties and candidates using social media, and attend offline events to which they were invited online. The researchers suggest that lower-threshold forms of political engagement on social media do not distract from higher-threshold activities, but are in fact strongly associated with them (Vaccari et al. 2015).

Social media can also function as a link to civic engagement. In their study of Western European local governments, Bonsón et al. (2015) find that media genres and content such as Facebook affect engagement. Topics related to municipal management are found to be of interest to citizens, so local governments need to devise ways to provide such information. They find that photographs are also beneficial for enhancing participation, which leads to engagement. Facebook is found to be effective because the platform functions have the potential to allow citizens to interact with the government through posting on government Facebook pages. The findings by Bonsón et al. (2015) indicate how social media is allowing citizens to become more empowered and more engaged in local issues.

Hofman et al. (2013) has also shown evidence of how Facebook can be useful for local governments in Germany because the platform provides timely information for citizens. Social media can be a useful communication channel for interacting with citizens and other stakeholders. Hofman et al. (2013) also agree that pictures and videos are beneficial for increasing engagement and that Facebook can create new opportunities for more communication behavior. Mossberger et al. (2013) have pointed out how social media has allowed for many of the major US municipal governments to become more open through the adoption of Facebook and other interactive platforms. Being more open has led to more interactivity, and has suggested how these technologies are allowing for better engagement.

Sobaci (2016) has created an inventory on how social media can be beneficial for local governments. They can improve (a) efficiency, (b) productivity, (c) local public services, (d) policy-making, (e) the strength of local democracy, (f) collaboration, and (g) the management of knowledge. However, risks accompanying social media for local governments include, for instance, (a) resources, (b) legal issues, (c) security, (d) information and content concerns, and (e) reputation management.

Political Participation and Civic Participation Through SNS in Japan

In Japan, municipalities are established in a two-tier structure: the *todoufuken* regional tier of prefectural or metropolitan governments and the *shikuchouson* local tier of cities, wards, towns, and villages. The partial existence of regions with jurisdictional disputes notwithstanding, Japan is divided into local *shikuchouson*, whose boundaries are incorporated into the shape of the regional *todoufuken* that contain them. Today, there are 47 *todoufuken*, containing 1,718 *shikuchouson*.

Moreover, the Japanese government is characterized by its relative "smallness of scale" internationally. Maeda (2014) finds that the number of state and local civil servants in Japan is quite small by carrying out an international comparison utilizing various sources of data such as OECD data, Nomura research data, National Tax Agency data, ILO data, and an international comparison of the number of personnel in the public sector for every 1,000 people; this information was made public by the Personnel Affairs Department of the Cabinet Office. Maeda calls the Japanese government "the state that does not employ its citizens," and suggests that this also inhibits the social progress of gender inclusion. The fact that the number of civil servants is low does not mean, however, that the number of activities conducted by the Japanese administration is small. The Japanese bureaucracy engages in a large number of activities through mobilizing the maximum amount of various resources that exist in society. Specifically, through use of the resources linked to networks such as industry groups, a large number of activities have been maintained with a small amount of resources. The Japanese political scientist and scholar of public administration Muramatsu has referred to this sociopolitical state as "a maximum mobilization system" (Muramatsu 1994). To put it another way, the political and social groundwork for political and public participation has already existed in Japan.

In addition to these characteristics, since the 1990s, the Japanese government has not only promoted efforts toward electronic governance, but it has also encouraged the use of the Internet and social media by local government, and has contributed to the promotion of indirect political and public participation. Prior efforts regarding electronic governance in Japan, included the promotion of the automation of large-scale routine tasks in the 1950s. In the late 1990s amid the spread of the Internet, "Reforming the General Plan for Promoting the Computerization of

Governance" (a cabinet decision dated December 20, 1997) stated that "as well as using the Internet and websites to provide online materials reporting the information that is made public every day, [including] every kind of administrative news that is important to the life of the people, we can also promote ever more completeness and timeliness of the material provided." In other words, each government ministry was encouraged to put in place a clearing system for informing the public about administrative news. Aside from this, in "Reforming the General Plan for Promoting the Computerization of Governance," basic provisions aimed at fostering the computerization of governance were adopted, meant to "promote the deployment of one computer for each person among the necessary personnel." Moreover, since the 2000s, with the aim of simplifying and optimizing administrative management to an even greater degree, efforts have been promoted regarding the following four pillars (MIC 2013: 180–185):

1. Promoting Internet use and moving administrative procedures online (such as the "e-Japan strategy"),
2. Optimizing tasks and systems,
3. Implementing cloud information systems,
4. Establishing and strengthening ICT governance in the administration.

To penetrate e-governance and make civic engagement or political participation more active, the role of local governments that provide public services close to citizens is quite strenuous. In addition, local governments have high expectations for e-governance because they have a strong preference for providing public services more effectively, increasing administrative efficiency, revitalizing regional industries by e-government, and so forth. The use of Internet-based functions is not new in Japan, as many municipalities are using e-mail newsletters for information dissemination such as safety alerts and weather advisories. According to the Ministry of Internal Affairs and Communications (MIC), as of April 2014, 93.6 percent of larger administrative structures such as large cities and prefectures in Japan distribute e-mail newsletters, and 22.8 percent of smaller administrative structures (small cities, towns, and villages) utilize this function for one-way information distribution (MIC 2014). According to survey results by the Ministry of Internal Affairs and Communications in 2013, 35.2 percent of the local governments in Japan make use of websites; 28.3 percent use a commercial SNS (i.e., Facebook); and 29.1 percent use blogs or micro-blogs (i.e., Twitter) for communication (MIC 2013). With regard to their reasons for using social media, 86.5 percent use social media for communicating to residents and local businesses; 72.7 percent for disseminating information to non-residents; and 58 percent have started using social media in case of disasters. Furthermore, the use of social networking services to promote municipalities can be observed in Japan, similar to examples in other nations such as Spain (Kiss 2015).

In Japan, in the past many local governments and Non Profit Organizations (NPO) have already attempted to promote public–private collaboration by launching bulletin board services and have also built customized systems for local social networking services that were appropriately named *Chiiki SNS* (local SNS), as well as utilized existing SNS platforms such as Facebook and Twitter (Shoji 2012; Noguchi and Ito 2013). Social experiments for operating bulletin board systems or *Chiiki SNS* (regional social networking services or systems) have been conducted by numerous Japanese local governments and non-profits in the past. For instance, the MIC conducted experiments deploying regional social networking systems in Chiyoda-ku, Tokyo, and Nagaoka City, Niigata Prefecture from, 2005 to 2006. Through these experiments, the notion of regional SNS systems has become familiar across Japan where regional networks have been established through these SNSs. Through them, the production, distribution and storage of regional information and their use for community development spread across the country in the 2000s (Shoji 2012, p. 68).

However, local governments are facing various challenges when they employ these social networking services. Nakano (2014) has found that local governments stop using their *Chiiki SNS* for three main reasons: (1) project re-evaluation; (2) an introduction of a time limit for usage; and (3) competition with other existing and nascent information and communications technologies (ICT). Furthermore, Kogawa (2012) studied the Fujisawa Citizens' e-Conference Room (also known as the *Commutto Fujisawa*), which was operated jointly by the Fujisawa municipal government and Keio University. They discovered that, even when business was stable, it remained difficult to fully outsource its operation to private parties. Employees of the Fujisawa municipal government felt strong pressure to maintain the SNS, which they considered a centerpiece business of Fujisawa City, and, consequently, they were reluctant to carry out reforms that would be seen as "irresponsible." In recent years, more and more local governments and advocates for public–private collaboration have begun to use existing social networking services such as Facebook and Twitter instead of bulletin board systems or *Chiiki SNS* (Shoji 2012, p. 56), and such use is expected to increase further in the future.

One noteworthy example in Japan is the case of Takeo City of Saga Prefecture, located in southern Japan. In recent years Takeo City has switched from using traditional websites to featuring its Facebook page as its main Internet presence. Takeo City has a history of progressive use of social media such as Twitter. For example, by 2010, it had provided Twitter accounts to a majority of its workers and has been encouraging them to post tweets, contrary to the policies of many corporations which prohibit the use of Twitter during working hours. The local government's aim was to promote the flow of ordinary information from the government so that citizens could feel closer to their municipality. During the flooding that occurred in the Takeo area on June 12, 2011, the mayor and other workers continued to distribute disaster information about road blocks and flooding, and subsequently were able to raise over one million yen in donations in part thanks to communicating their relief efforts via Twitter. Their success in using Twitter evolved into a strategy for further enhancing awareness of the workings of the municipal government, and eventually they expanded their online presence to the use of Facebook.

Takeo City provides information on services for residents, including in areas such as child-rearing support, general safety, tourist information, and business or procurement information. Using inline frames, the municipality embeds local government information in Facebook, but it also uploads such pages onto a server outside of Facebook to avert risk in case of service failure or changes in Facebook policy. Their Facebook page is set to open access for those who do not have Facebook accounts. Communication is facilitated between the government and Facebook account holders who can comment on and "like" any updates posted by the city.

Using Facebook has certain merits for local governments. It allows for quicker service, increased accountability, lower IT server maintenance costs, and greater opportunities for interactivity among citizens and the government. However, there are also a number of potential disadvantages: Facebook implementation may be intimidating to citizens with varying levels of digital skills (IT literacy or computer literacy). Furthermore, to ensure fairness to all citizens regardless of their levels of digital literacy, local governments must also consider balancing information provision to address potentially different levels of interest in civic activities among citizens.

There is also a certain degree of risk in the case of Facebook, as some services may be phased out and policies concerning privacy and personal information may be changed. Such concerns about platform durability may inhibit new users from joining the service. In Japan, where government polls have identified the potential for misuse of personal information to be the most important concern among users overall, entrusting their shared personal or private information to Facebook is a particularly relevant issue. Despite the low risk, some entries are potentially subject to trolling by other Facebook users (MIC 2013).

In addition to the case study of Takeo City, Joyo ARC, a think-tank based in Japan's Ibaraki Prefecture, conducted a survey concerning social media usage by local governments in the northeastern Kanto area covering Ibaraki Prefecture, Tochigi Prefecture and Gunma Prefecture. The study found that 18 local governments were operating official social media accounts in Ibaraki Prefecture, and nine local governments such as Mito City, Tsukuba City, and Hitachi City operated multiple social media accounts in 2012. They also found that nine local governments in Tochigi Prefecture, including Utsunomiya City and Oyama City, operate Twitter accounts and blogs. Social media such as Twitter or Facebook accounts were also being used by five local governments in Gunma Prefecture such as Numata City, Ooizumi City, and Tsumagoi City (Joyo ARC 2012).

According to Joyo ARC, a noteworthy example was the case of Tsukuba City of Ibaraki Prefecture. As mentioned in Chap. 2, Joyo ARC pointed out how the Tsukuba municipal government began to use Twitter to enhance communication between city hall workers and residents in January 2011. Immediately after it started using Twitter, Tsukuba City faced serious damage caused from the Great East Japan earthquake on March 11, 2011. Many citizens rushed to call Tsukuba City Hall directly, view the city's webpage, or listen to community radio to obtain disaster-related information. However, in the midst of the flood of inquiries via telephone and the Internet, this basic infrastructure was not functioning as expected. As documented in Chap. 2, Joyo ARC's report mentions how remarkable it was that the Tsukuba municipal workers started using Twitter to provide and collect disaster information as efficiently as possible. They retweeted the tweets by the residents when they thought they had found information that was worth believing. In the days following the Fukushima Daiichi nuclear power plant accident, the citizens' voices that were collected via Twitter were not always positive with regard to the Tsukuba municipal workers. For example, there were a number of critical messages on Twitter when they could not accept all the disaster victims from Fukushima Prefecture immediately. However, there were many supportive messages for the city workers on Twitter, and overall these positive messages and tweets encouraged the municipal workers during the period when Tsukuba City Hall was reconstructed (Joyo ARC 2012). Joyo ARC (2012) confirms the findings and suggestions stated in Chap. 2.

In viewing both the pros and cons of implementing Facebook services, many Japanese local municipalities have determined that the benefits outweigh the risks of incorporating Facebook usage into their daily operations. They have initiated, or are currently planning to create, Facebook pages to provide fora for citizens and government so that community members can exchange ideas and participate in social activities (MIC 2013). However, we have relatively little understanding of how social networking sites such as Facebook are being used by local governments to create a more collaborative relationship among citizens and promote citizen engagement, and what policy areas tend to get more attention and achieve more engagement by citizens.

Objective of the Study in This Chapter

This study investigates how local governments in Japan use Facebook. Our aim is to examine the specific areas of policy that each local government Facebook page focuses on. We posit that some Facebook pages focusing on some specific policy areas are more likely to have a large number of fans, and that other Facebook pages that focus on other areas of policy are more likely to have more engagement. Our rationale behind this assumption is based on the fact that the level of civic engagement could differ depending on the area of policy involved. As mentioned earlier, some policy areas are expected to have higher engagement, while others are not. Therefore this study investigates the following three research questions.

- RQ1: What is the major focus of local governments in Japan when they use Facebook?
- RQ2: What policy area focused on by a local government Facebook page accumulates more fans?
- RQ3: What policy area focused on by a local government Facebook page accumulates more engagement?

Methodology

To analyze the focus of each local governments' Facebook page, we used a list provided by the *Chiiki SNS Kenkyukai* which has a "list of local governments' Facebook page in Japan (as of March, 2013)." Specifically, the list is an inventory of the Facebook pages that are associated with the official webpages of: (1) prefectural governments; (2) municipalities in prefectural capital cities and designated cities (the redundant pages with prefectural governments were removed); (3) Tokyo's 23 special wards, and pages posted on news sites. A total of 466 Facebook pages are listed. We removed 38 pages that were defunct and three pages that were being operated by individuals, not governments. Through this process, we had a remaining total of 425 Facebook pages (see Appendix A).

The Facebook profiles of the 425 pages were extracted and analyzed statistically. The profile data analyzed in this study included: (1) the total number of people who "liked" the page (fans); (2) people talking about this page (engagement); (3) brief information of the page; and (4) a more concrete description of the profile. By using the Blockspring application, data was extracted from the 425 Facebook pages between July 7, 2016 and July 15, 2016.

Data Analysis Method

The area of policy focused on by each local government's Facebook page was analyzed through the KH Coder text mining software. The "Brief information of the page" of each Facebook profile was analyzed using KH Coder that allows for analysis of Japanese-language text along with Chinese-, Korean- and Russian-language texts. KH Coder conducts morphological analysis for the Japanese language utilizing the ChaSen morphological analysis tool developed by the Nara Institute of Science and Technology. This tool can be employed to assist in conducting computer based content analysis of Japanese text. We used this software for determining co-occurrence patterns in the paragraphs of the text provided in the Facebook profiles. The text is analyzed based on the following rules to enhance the accuracy of analyzing Japanese.

1. The words with inflections are taken out after they are changed to the basic forms.
2. The general words that are used in any sentences such as postpositional particles and auxiliary verbs are omitted from the target of analysis.

Only "nouns" and "*suru* verbs" (verb formed by adding *suru* to a noun in Japanese) are counted because our main focus is to discover the policy areas that local governments are focusing on. As a result, the remaining data was extracted and the co-occurrence network (modularity) is illustrated in the following section.

Results: Text Frequency and Co-occurrence Network

For analysis of each local governments' profile on its Facebook page, the "nouns" and "suru verbs" (verb formed by adding "suru" to a noun) were measured. Descriptive statistics on frequencies of "nouns"(名詞) and "suru verbs"(サ変名詞) are shown in Table 3.1. With regards to nouns, the word "information" (情報) is the most frequent word (184) with "page"(ページ), "municipal halls"(市役所), "events"(イベント), "communities"(地域), "attractive"(魅力), "ward offices"(区役所), "center"(センター), "projects" (プロジェクト) and "accounts"(アカウント) following. Regarding "suru verbs," "transmit"(発信) is the most frequent word, with "sightseeing"(観光), "management"(運営), "public relations"(広報), "introduce"(紹介), "promotion"(進行), "propulsion"(推進), "notice"(お知らせ), "activities" (活動), and "support" (支援) following.

Figure 3.1 illustrates the co-occurrence network of Facebook profiles provided by the analysis results of KH Coder. Nodes are displayed larger when frequencies are higher, and edges (lines between the bubbles) are denoted thicker when co-occurrences of the pair of the adjacent nodes are more frequent. It is immediately visible that at the right bottom side of the co-occurrence network, we have community-development or community-promotion-related words such as "community," "events," "introduce," "attractive," "promotion" etc. With regards to "community," there are co-occurrence relationships among "promotion," "information," and "introduce," and at the same time, each text co-occurs among "events," "local specialty," and "transmit". These characteristics of the co-occurrence network revealed that "community development/community promotion" is the main focus of public policy of the local government.

At the same time, we can observe (a) agriculture, (b) environment, (c) enterprise, (d) civic engagement, (e) employment, (f) (agriculture) forestry and fisheries, (g) industrial development, (h) sports, (i) childcare (j) art and culture, (k) tourism and (l) public relations as policy areas.

Table 3.1 Analysis results of frequency of text

	Nouns	N	Suru verbs	N		Nouns	N	Suru verbs	N
1	情報	184	発信	64	26	県民	12	共同	7
2	ページ	154	観光	60	27	スポーツ	10	防災	7
3	市役所	49	運営	39	28	県庁	10	離島	7
4	イベント	47	広報	35	29	市立	10	PR	6
5	地域	38	紹介	31	30	農業	10	安心	6
6	魅力	36	振興	29	31	ネット	9	管理	6
7	区役所	32	推進	29	32	芸術	9	施設	6
8	センター	27	お知らせ	23	33	事務所	9	試験	6
9	プロジェクト	24	活動	22	34	王国	8	消費	6
10	アカウント	22	支援	21	35	環境	8	創造	6
11	職員	22	採用	17	36	森林	8	促進	6
12	フェイス	21	生活	15	37	団体	8	提供	6
13	ブック	21	掲載	14	38	特産	8	連携	6
14	ホームページ	21	応援	12	39	農林	8	お願い	5
15	県立	21	交流	12	40	ネットワーク	7	オープン	5
16	サイト	19	子育て	12	41	活性	7	コメント	5
17	行政	17	お伝え	11	42	社会	7	位置	5
18	産業	17	協議	11	43	男女	7	移住	5
19	文化	17	開催	10	44	動物	7	運用	5
20	協会	14	対策	10	45	美術館	7	雇用	5
21	事業	14	参画	8	46	福祉	7	就職	5
22	市民	13	総合	8	47	本部	7	選挙	5
23	都市	13	ガイド	7	48	未来	7	配信	5
24	委員	12	関連	7	49	目的	7	復興	5
25	企業	12	企画	7	50	ご覧	6	了承	5

Analytical Results of Policy Area

The data shown in Table 3.2 indicate the analytical results of the policy area focused on each by Japanese local government. Community development is the most frequent (157), followed by tourism (53), entertainment (29), childcare; education and learning (25), public relations (24), agriculture, forestry, and fisheries (21) and employment (15). On the other hand, fewer than ten local governments focused on welfare-related issues such as animal welfare (4) and social welfare (4); safety-related issues such as public safety (3) and consumer safety (2); and diversity-related issues such as gender equality (5), multicultural coexistence (3), and international relations (2). At the same time, only six local governments focused on civic engagement.

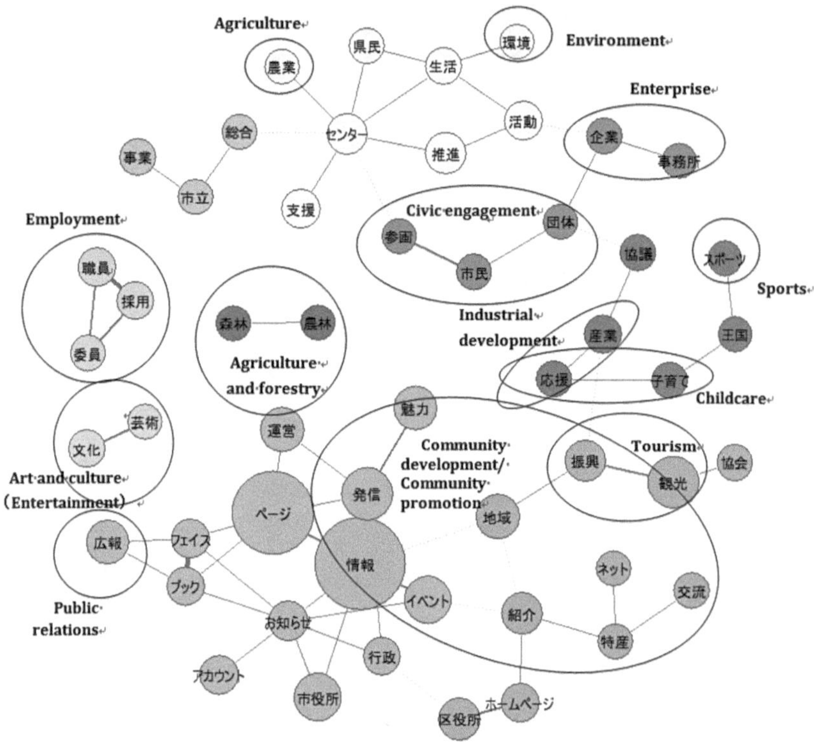

Fig. 3.1 Co-occurrence patterns of Facebook profiles (modularity, minimum frequency 8)

Subsequently, we examined which area of policy would increase fans or engagement of a Facebook page of a Japanese local government. We first analyzed the total number of fans of each local government's Facebook page. By the following classification created for assisting with better interpretation of the result of this analysis, 46.8 percent are small-scale (under 1,000 fans); 37.2 percent are mid-scale (1,001–5,000 fans); and 16 percent are large-scale (over 5,001 fans). Through further inspection of the percentage of various numbers of fans per page, those with 1,001–2,000 fans was the highest (at 19.3 percent), followed by 501–1,000 fans (at 18.8 percent), and those with under 250 fans (at 15.1 percent) and 251–500 fans (at 12.9 percent). The percentage of pages with over 9,000 fans was 7.3 percent. Therefore, through our analysis, we can conclude that most

Table 3.2 Policy area

Areas of policy	%	*Count*
Community development	36.9	157
Tourism	12.5	53
Entertainment	6.8	29
Childcare, education, and learning	5.9	25
Public relations	5.6	24
Agriculture, forestry, and fisheries	4.9	21
Employment	3.5	15
Environment	3.3	14
Health care and sports	2.6	11
Land, infrastructure, and transport	2.6	11
Disaster prevention	2.4	10
Foods	2.1	9
Enterprise	1.6	7
Industrial development	1.6	7
Civic engagement	1.4	6
Gender equality	1.2	5
Animal welfare	0.9	4
Social welfare	0.9	4
Election	0.7	3
Multicultural coexistence	0.7	3
Public safety	0.7	3
Consumer safety	0.5	2
International relations	0.5	2
Total	100.0	425

Japanese local government Facebook pages have fewer than 1,000 fans, and their fan base is usually small-scale; however, the number of Japanese local government Facebook pages with over 9,000 fans is noteworthy.

We next analyzed each policy area (The aggregate data of all policy areas is in Appendix B.) First, among the policy areas of tourism, public relations, and community development, these pages have the highest-level median in seven policy areas. Tourism and public-relations-related pages have over 15 percent among the large-scale (over 9,000 fans) fan base pages; therefore, these two policy areas have the largest fan base. Second, with regard to agriculture, forestry, and fisheries; childcare; education and learning; and civic engagement, these pages have a moderate-level median in seven policy areas. Among these policy areas, about 70 percent of the pages have fewer than 1,000 fans. Third, with regard to employment, the pages have the lowest-level median in seven policy areas. All the pages focused on employment have fewer than 1,000 fans.

We next analyzed the distribution of the degree of engagement, and we observed a fairly even distribution of engagement from 0 to over 1,000. We also examined the engagement of each area of policy. For better assessment of the distribution, and we found it was divided into three groups: (1) low engagement, (2) mid-scale engagement, and (3) high engagement. Employment; childcare; education and learning; civic engagement; and agriculture, forestry, and fisheries have low engagement, although public relations and community development have relatively high engagement, and tourism has mid-scale engagement.

Discussion

This study investigated how local governments in Japan use Facebook. We examined the specific policy areas focused on by local governments in Facebook and found some to have more fans, and others to have more engagement. This study investigated the following three research questions.

RQ1: What is the major focus of local governments in Japan when they use Facebook?
RQ2: What policy area focused on by a local government Facebook page accumulates more fans?
RQ3: What policy area focused on by a local government Facebook page accumulates more engagement?

Our study found that Japanese local governments use Facebook for (a) agriculture; (b) environment; (c) enterprise; (d) civic engagement; (e) employment; (f) (agriculture), forestry and fisheries; (g) industrial development; (h) sports; (i) childcare; (j) art and culture; (k) tourism; and (l) public relations as their focused policy areas. In answering RQ1, among these policy areas, community development and community promotion are the most common policy areas, followed by tourism; entertainment; childcare; education and learning; public relations; and agriculture, forestry, and fisheries. By contrast, the policy areas there was little evidence with regard to diversity or social equality such as gender equality or multicultural coexistence. This result indicates that Japanese local governments do not at present operate social media with regard to these policy areas, and, as a result, are lowering the civic engagement or political participation via Facebook that has been achieved among Japanese local governments.

With regard to RQ2, which makes an inquiry into which area of policy tends to accumulate more fans, employment was found to have a small fan base. By contrast, tourism, public relations and community development have a large-scale fan base and agriculture, forestry and fisheries, childcare, education and learning and civic engagement all have a mid-scale fan base. With regard to RQ3, which considers which areas of policy tend to accumulate more engagement, we found that employment, childcare, education and learning and civic engagement have low engagement, tourism has mid-scale engagement, and public relations and community development have higher engagement.

Figure 3.2 illustrates this study's findings regarding how the number of followers and the level of engagement fluctuate depending on each policy area. We discovered four different policy areas based on the number of followers and the level of engagement. The first group has a high number of followers as well as level of engagement. Public relations and community development are categorized in this group. With regard to this group, the function of the local governments' Facebook pages would be not only to provide and share information, but also to create opportunities to enhance communication and engagement among citizens. The second group has a moderate number of followers, while engagement is relatively low. Childcare; education and learning; civic engagement; and agriculture,

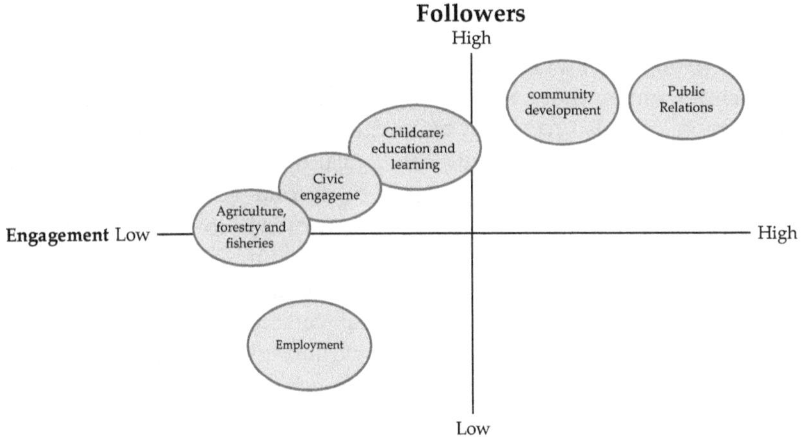

Fig. 3.2 Fans × engagement × policy area

forestry, and fisheries are categorized in this group. With regard to this group, the current function of local governments' Facebook pages is basically confined to providing and sharing information, and more effort could be required to make civic engagement via Facebook more proactive. The third group has a low number of followers as well as low engagement via Facebook. Employment is categorized in this group. We can also suggest that the number of fans or followers does not influence the level of engagement. The fourth group could consist of obscure policy areas in Japan, which Japanese local governments do not deal with very much through Facebook. Policy areas with regard to diversity or social equality such as gender equality or multicultural coexistence are categorized in this group.

Establishing collaboration and enhancing civic engagement is an important aspect in many areas of society; however, collaboration among citizens, the public sector, and the private sector is one of the most important issues for many industrialized nations. Social media such as Facebook can provide a possible path for local governments to become more open and allow for such collaboration. In this scenario, citizens can be more involved in public service and policy decisions.

Our study finds that Japanese local governments use Facebook mainly for community development and community promotion, while policy with regard to diversity or social equality such as gender equality or multicultural coexistence rarely exists. In addition, we categorized four different policy areas based on the number of followers and level of engagement. The first group has a high number of followers as well as high engagement in which the function of Facebook would not only be to provide and share information, but also to create opportunities to enhance communication and engagement among citizens. Public relations and community development are categorized in this group. The second group has a moderate number of followers and engagement is relatively low; the function of Facebook would be basically confined to providing and sharing information at that stage, and greater effort could be required to make civic engagement via Facebook more proactive. Childcare; education and learning; civic engagement; and agriculture, forestry, and fisheries are categorized in this group. The third group has a low number of followers as well as low engagement via Facebook. Employment is categorized in this group. The final group consists of obscure policy areas that Japanese local governments do not deal with much through Facebook such as diversity, social equality, gender equality, or multicultural coexistence.

The findings described in this chapter focus solely on Japanese local government Facebook pages, but the wider implications suggest a framework for assessing how local governments can promote political participation and civic engagement more efficiently.

References

Bonsón, E., Royo, S., & Ratkai, M. (2015). Citizens' engagement on local governments' Facebook sites: An empirical analysis: The impact of different media and content types in Western Europe. *Government Information Quarterly, 32*(1), 52–62.

Boulianne, S. (2015). Social media use and participation: A meta-analysis of current research. *Information, Communication & Society, 18*(5), 524–538.

Eto, M. (1998). Jichitai fukushi to shimin sanka [Citizens' participation in municipal welfare policy process]. *The Quarterly of Social Security Research, 33*(4), 403–413.

Fisher, R. D., Svendsen, S. E., & Connolly, J. (2015). *Urban environmental stewardship and civic engagement: How planting trees strengthens the roots of democracy.* London: Routledge.

Gil de Zúñiga, H., Jund, N., & Valenzuela, S. (2012). Social media use for news and individuals' social capital, civic engagement and political participation. *Journal of Computer-Mediated Communication, 17*(3), 319–336.

Hofmann, S., Beverungen, D., Rackers, M., & Becker, J. (2013). What makes local governments' online communications successful? Insight from a multi-method analysis of Facebook. *Government Information Quarterly, 30*(4), 387–396.

Huntington, P. S., & Nelson, M. J. (1976). *No easy choice.* Harvard: Harvard University Press.

Ikeda, K., Kobayashi, T., & Hoshimoto, M. (2008). Does political participation make a difference? The relationship between political choice, civic engagement and political efficacy. *Electoral Studies, 27*(1), 77–88.

Jennings, M. K., & Zeitner, V. (2003). Internet use and civic engagement – A longitudinal analysis. *Public Opinion Quarterly, 67*(3), 311–334.

Joyo ARC. (2012). Sōsharu medhia o katsuyou shita machi dukuri no kanousei: Jichitai, chiiki jūmin no torikumi [City development possibilities through social media: Neighborhood associations and resident activities]. *Joyo ARC, 44*(511), 16–33.

Kabashima, I. (1988). *Seiji sanka [Political participation].* Tokyo: Tokyo University Press.

Kavanaugh, A. L., Fox, E. A., Sheetz, S. D., Yang, S., Li, L. T., Shoemaker, D. J., Natsev, A., & Xie, L. (2012). Social media use by government: From the routine to the critical. *Government Information Quarterly, 29*(4), 480–491.

Kiss, J. (2015). Welcome to Jun, the town that ditched bureaucracy to run on Twitter. *The Guardian* (online). Retrieved 7 Aug 2015, from http://www.theguardian.com/technology/2015/jul/02/twitter-jun-spain-bureaucracy-local-government

Kobayashi, T., & Ikeda, K. (2005). Mou hitotsu no debaido: Keitai debaido no sonzai to sono kiketsu. [Another divide: The existence of Mobile phone divides]. In K. Ikeda (Ed.), *Intānetto komyunityī to nichijō sekai* [Everyday world and Internet community] (pp. 47–66). Tokyo: Seishin Shobo.

Kodagiri, Y. (2014). *Gyōsei-shiminkan kyōdō no kouyou: Jisshouteki sekkin* [Cooperation between administration and citizens: An empirical approach]. Tokyo: Horitsu Bunkasha.

Kogawa, I. (2012). Fujisawashi shimin denshi kaigishitsu 15 nen no rekishi: Sono doutai kara nani wo manabuka [The 15 years of the Fujisawa citizen electronic meetings: What can we learn]. *Journal of the Musashi Sociological Society, 14*(1), 35–54.

Kushin, J. M., & Yamamoto, M. (2010). Did social media really matter? College students' use of online media and political decision making in the 2008 election. *Mass Communication & Society, 13*(5), 608–630.

Lee, G., & Kwak, Y. H. (2012). An open government maturity model for social media-based public engagement. *Government Information Quarterly, 29*(4), 492–503.

Linders, D. (2012). From e-government to we-government: Defining a typology for citizen coproduction in the age of social media. *Government Information Quarterly, 29*(4), 446–454.

Livingstone, S., & Markham, T. (2008). The contribution of media consumption to civic participation. *British Journal of Sociology, 59*(2), 351–371.

Maeda, K. (2014). *Shimin o yatowanai kokka: Nihon ga koumuin no sukunai kuni heto itatta michi* [A state without civil servants: Japan's public sector in comparative perspective]. Tokyo: Tokyo University Press.

McAtee, A., & Wolak, J. (2011). Why people decide to participate in state politics. *Political Research Quarterly, 64*(1), 45–58. doi:10.1177/1065912909343581.

Ministry of Internal Affairs and Communication. (2013). *2013 white paper: Information and communication in Japan*. Tokyo: Ministry of Internal Affairs and Communication.

Ministry of Internal Affairs and Communications, Japan (MIC). (2014). *Chihōjichi Jōhōkanri Gaiyō (Denshi Jichitai no Suishin Jōkyō Heisei 26-nen 4-gatsu 1nichi genzai)* [Local Government Information Overview] (Implementation Status of E-Municipalities from April 1, 2014 to Present) (p. 17). Tokyo: Ministry of Internal Affairs and Communications. Retrieved 11 Aug 2015, from http://www.soumu.go.jp/denshijiti/060213_02.html

Mossberger, K., Wu, Y., & Crawford, J. (2013). Connecting citizens and local governments? Social media and interactivity in major U.S. cities. *Government Information Quarterly, 30*(4), 351–358.

Muramatsu, M. (1994). *Nihon no gyōsei: Katsudou gata kanryōsei no henbou* [Japanese administration: The change of active bureaucracy]. Tokyo: Chuko Shinsho.

Nakano, K. (2014). Kansei chiiki SNS ga haishi ni itaru keii ni kansuru kousatsu: Jichitai shokuin heno intabyū chōsa yori [Why the Chiiki SNS were abolished: Interviews with local government employees]. *Social and Economic Systems Studies, 35*, 35–42.

Nishio, M. (2004). Hashigaki [Foreword]. In T. Nishio (Ed.), *Jūmin, komyunityī tono kyōdō*. [Cooperative work of residents and community] (p. iii). Tokyo: Gyousei.

Noguchi, M. & Ito, N. (2013). Jichitai ni okeru Facebook kouhou ni kansuru medhia kouka sokutei: Sagaken takeoshi no sōsharu kyapitaru to shibikku pawā heno eikyō. [Measurement of media effects of Facebook PR by local governments: Effects on civic poer and social capital in Takeo, Saga.] *Journal of the Japan Information-Culture Society, 20*(2), 35–42.

Pateman, C. (1970). *Participation and democratic theory*. Cambridge: Cambridge University Press.

Rochon, T. R., & Kabashima, I. (1998). Movement and aftermath: Mobilization of the African American electorate, 1952–1992. In J. Geer (Ed.), *Politicians and party politics* (pp. 102–121). Baltimore: Johns Hopkins University Press.

Shah, D. V., McLeod, J. M., & Yoon, S. H. (2001). Communication, context, and community – An exploration of print, broadcast, and Internet influences. *Communication Research, 28*(4), 464–506.

Shah, D., Cho, J., Eveland, W., & Kwak, N. (2005). Information and expression in a digital age – Modeling Internet effects on civic participation. *Communication Research, 32*(5), 531–565.

Shoji, M. (2012). Chiiki ni okeru shakai nettowāku to jōhō tsūshin gijutsu [Information and communication technology and social networks in regions]. In S. Sugimoto (Ed.), *Jōhōka jidai no rōkaru komyunityī: ICT o katsuyou shita chiiki netowāku no kouchiku* [Local communities in the age of information: Construction of local networks through ICT] (pp. 61–80). Osaka: National Museum of Ethnology.

Sobaci, M. Z. (Ed.). (2016). *Social media and local government: Theory and practice*. London: Springer.

Takahashi, H., & Sato, T. (Eds.). (2013). *Shinsetsu shimin sanka kaiteiban. [Citizen participation new and revised]*. Tokyo: Kojinsha.

Vaccari, C., Valeriani, A., Barbera, P., Bonneau, R., Jost, J. T., Nagler, J., & Tucker, J. A. (2015). Political expression and action on social media: Exploring the relationship between lower and higher-threshold political activities among Twitter users in Italy. *Journal of Computer-Mediated Communication, 20*(2), 1–19.

Vissers, S., & Stolle, D. (2014). The Internet and new modes of political participation: Online versus offline participation. *Information, Communication & Society, 17*(8), 937–955.

Wellman, B., & Haythornthwaite, C. A. (Eds.). (2002). *The Internet in everyday life*. Malden: Blackwell Pub.

Xenos, M., Vromen, A., & Loader, B. D. (2014). The great equalizer? Patterns of social media use and youth political engagement in three advanced democracies. *Information, Communication & Society, 17*(2), 151–167.

Young, D. R. (2000). Alternative models of Government-Nonprofit sector relations: Theoretical and international perspectives. *Nonprofit and Voluntary Sector Quarterly, 29*(1), 149–172.

Zukin, C., Keeter, S., Andolina, M., Jenkins, K., & Delli Carpini, X. M. (2006). *A new engagement? Political participation, civic life, and changing American citizen*. Oxford: Oxford University Press.

Sae Okura is an assistant professor in the Faculty of Humanities, Law and Economics at Mie University in Japan. She conducts research in politics and social policy studies. Her recent research interests focus on political participation by minority groups such as the disability community, sexual minorities, and single-parent families. She analyzes quantitative survey data collected in Japan.

Muneo Kaigo is an associate professor in the Faculty of Humanities and Social Sciences at the University of Tsukuba in Japan. He teaches courses in media communication in the Graduate School of Humanities and Social Sciences, media management in the Graduate School of Business Sciences and communication science in the College of Comparative Culture at the University of Tsukuba. He is currently leading a joint research project on social media uses among local municipalities in Japan with cooperation from the municipal government of Tsukuba and Intel Corporation Japan.

CHAPTER 4

Civil Society, Social Media and Facebook Usage by Local Governments: Birth of the Tsukuba Study and the Tsukuba Civic Activities Cyber-Square

Muneo Kaigo and Leslie Tkach-Kawasaki

The Need for Connecting Citizens and Local Governments in Japan

This chapter analyzes a case of social media usage by a local municipality in Japan, focusing on the possibilities and problems of complementary communication channels such as social networking services. Rapid transmission allows social media to facilitate communication activities in new ways and offers new avenues for information interaction between citizens and local governments, as we have seen with the example in Chap. 2 of the microblog Twitter. Twitter can connect citizen followers through a local government's Twitter account and citizens can receive updates continuously about public services and events.

M. Kaigo (✉) • L. Tkach-Kawasaki
Faculty of Humanities and Social Sciences, University of Tsukuba, Tsukuba, Ibaraki, Japan

M. Kaigo (ed.), *Social Media and Civil Society in Japan*, DOI 10.1007/978-981-10-5095-4_4

Other SNS platforms, such as Facebook, can also allow citizens to experience more personal connections with their local governments. These types of online relationships demonstrate not powerfully committed personal bonds, but weak links (Resnick et al. 2012; Wellman et al. 2003) that acknowledge the existence of a connection without requiring a strong bond among people using social media. If strong bonds exist among users regardless of social media, then they can mobilize people, as in the case of the 2010–2011 Arab Spring political movements (Howard et al. 2011; Howard and Parks 2012). In general, just like websites, social media usage efficiently reduces distances between people and effectively connects members of modern society.

As indicated in Chap. 2, in the immediate aftermath of the Great East Japan Earthquake that occurred on March 11, 2011, voice communication through normal telephone lines was made difficult in the affected areas due to the initial congestion caused by the sudden increase of calls and broken lines. Immediately after the 1994 Northridge, CA earthquake, Pacific Bell was down in many affected areas, and the 1995 Awajishima-Hanshin earthquake left similar breakdowns in telecommunications lines followed by an overconcentration of incoming calls into the affected regions (Hooper 1999; Kroll-Smith 2001). Follow-up studies and reports have indicated that alternative ways of communication were effective for communications probing the safety of friends, relatives and close ones during these two earthquakes (Noam and Sato 1995; Kazuma 1995) to alleviate congestion due to concentration of phone calls into emergency areas. The Niigata-Chuetsu earthquake that hit the middle northern area of Japan on October 2004 disrupted telephone, cellular and Personal Handyphone System (PHS) communication by severing optic fiber lines and interrupting services, along with causing power failures.

As is in many natural disaster-related situations, people often became alarmed and nervous, needing reassurance that their loved ones were safe. Some needed to convey delays or cancellations resulting from disruptions to transport networks. Since the 1998 implementation of the Japanese disaster emergency voice message board (*dengon dial*) 171 operated by NTT, this has been regarded as the only alternative to normal telephone calls in such situations. However, in contrast to fixed telephone lines, Internet access through mobile devices was found to be relatively robust and resilient in comparison to the normal telecommunication channels. Voice over IP, SMS (Short Message Services) and email are quicker solutions for communication and overcoming telecommunication breakdowns or congestions in such disasters. Radio is also more effective due to its portability and its ability to function with batteries in comparison to televisions that are of no use during power outages. Social media such as Facebook and Twitter have proved more effective in disseminating information through the Internet than websites hosted on local servers which may have been put out of action as a result of the sort of power outages that occurred early in disasters such as the southern California wildfires and the Sichuan earthquake in 2008 (Sutton et al. 2008; Mills et al. 2009), or the hurricanes that hit the Atlantic coast of the USA in the same year (Hughes and Palen 2010). White (2010) has also discussed the role of social media in crisis-mapping during local-level disasters. Twitter was found to be most effective at quickly spreading information—whether true or false—in Japan during the Great East Japan Earthquake. In Chap. 2, I examined how Twitter has the potential to transmit false rumors very rapidly in Japan during disasters, therefore Twitter usage requires some further consideration as to how to avoid some of the unfavorable consequences that were observed during the Great East Japan Earthquake in 2011.

Following the Great East Japan Earthquake, many Japanese municipalities began to experiment with various social media other than Twitter, including Facebook. As power outages affected the local servers that were hosting Internet services in Japan, the Great East Japan Earthquake demonstrated how many of the expensive customized *Chiiki SNS* platforms were useless during times of unstable power supply, particularly as many areas were required to conserve energy via planned power outages during the weeks immediately after the Great East Japan Earthquake. Such situations were unimaginable before the disaster and served as catalysts to change the minds of conservative decision-makers within Japanese municipalities to outsource SNS platforms instead of constructing their own local platforms. During the Great East Japan Earthquake, researchers observed that through social media, a communication network of social support was formed in municipalities experiencing mid-level disruptions, and social interaction was made possible throughout a wide region (Hashimoto and Ohama 2014; Inoue 2013). Social media became an information conduit and knowledge source for the citizens of Japanese municipalities with mid-level disruptions in the early days after the disaster. This practical utilization suggests that this observed means of transmitting and receiving information is effective and potentially beneficial as a channel for municipal government communications.

Given this potential, in this chapter, we examine how the city of Tsukuba in Ibaraki prefecture experimented with the regular use of Facebook on an initial trial basis for a six-month period in early 2012 as a means of enhancing its citizens' social capital. This city's experience is rather unique, as during the experimentation period, it experienced its own local disaster when a tornado ripped through the north part of the city in early May 2012. Thus, the time period allows us to investigate the regular use of social media as a means of communicating local government information, as well as its use during a disaster situation. After examining the literature on social media, its relationship with social capital, and the background of social media and its use among municipalities in Japan, we look at the Tsukuba Civic Activities Cyber-Square experiment period and how the experiment was conducted by the City of Tsukuba in early 2012 and during the events of the May 6, 2012 tornado. In the final section of the chapter, we review the implications of the six-month experimental period and the use of social media during the tornado disaster, and contemplate the possibilities of social media utilization on the local government level in Japan to enhance the relationship between government and citizens.

Social Capital, Media, Design, Community and Civic Participation

As reviewed in Chap. 2, social capital is formed through the social contexts of trust, norms, and networks to make a more efficient society by promoting the resources that exist in the harmony among positive human relationships (Putnam 2000). Exchanges of ideas during times of difficulty build human relationships and demonstrate a means by which communication activities build social capital. Putnam (2000) has termed bridging and bonding to be the two ways of building social capital; however, when viewing this idea in the perspective of SNS-based communication activities, these two separate types of social capital may both be present. Communication that bonds close individuals may occur in SNS, but bridging among diverse individuals is equally possible with SNS; thus the intermix of the two types of building social capital needs to be considered as well.

Another approach directly related to the theme of this study is "social design" that explains the process of citizen involvement in community building and effective steps in tackling social issues in Japan (Kakei 2013). Social design is related to the Japanese concept of *machizukuri* (literally translated as "building community") as a means of encouraging local residents to participate in their local communities. As communication among residents is one of the core concepts, practical approaches—rather than theoretical constructs—are integral in developing and using online social spaces where residents can interact to tackle problems or issues.

The enormous economic growth and population increase in Japan during the twentieth century has now been replaced with the decline of both in recent years. This reversal of both economic and population growth has led local policy-makers to revise many previously uniform and standardized approaches to dealing with various social issues. Declines in tax revenues have led to cuts in budgets and a reduction in the number of government personnel. Management of social issues by the public sector alone is no longer possible in Japan, and this situation has offered opportunities for many citizen organizations and NPOs (non-profit organizations) to assume some of the roles that have traditionally been thought to be the responsibility of local Japanese governments. In this new fiscal environment, policy-makers need to focus more on constructing and designing social systems that take advantage of the available resources that enable them to accomplish their tasks, rather than creating new schemes or continuing to build new departments or structures. Many social actors share the same concerns towards society and each can play a role in building communities. Social design enables these social actors to communicate and share their concerns towards society and allow for coordination to actively use available resources and people (Kakei 2013). Many local governments in Japan are in need of such social design due to the increasing number of social issues that need to be addressed to help maintain local communities. One way that social design can be applied to local governments in Japan is to implement social media as a means of connecting civil society. As social media such as SNS can be used to virtually connect a new variety of social actors, social issues can be shared simultaneously through social media and all actors can begin participating and communicating with one another to address what needs to be done. Carey (1989) indicates how Internet communication should be viewed as a cultural transaction of meaning production. Online community message production in itself is a cycle of cultural life and recreates the many facets of a community's culture. Appadurai (1995) points out how locality is a "complex phenomenological quality, constituted by a series of links between the sense of social immediacy, the technologies of interactivity and the relativity of contexts" (Appadurai 1995, p. 204). In this sense one can stay in a set social arena or virtual neighborhood, as one can live in a spatial neighborhood and also virtual neighborhood that is defined by different rules in regards to boundaries (Pedersen 2003). Rheingold (1993) describes a "virtual community" as "social aggregations that emerge from the Net when enough people carry on ... public discussions long enough, with sufficient human feeling, to form webs of personal relationships in cyberspace" (Rheingold 1993, p. 5). Overall declines in social capital discussed by Putnam (2000) can be avoided in Japan by using social media in the process of redesigning the concept of "community."

With regard to previous research on social media increasing civic participation, McAtee and Wolak (2011) discuss social networks and motivation for participation in civil society. The emergence of a network society allows for everyday online communication and interconnectedness (Wellman and Haythornthwaite 2002). Virtual communities have the following shared traits of: (1) space, (2) shared practice, (3) shared resources and support, (4) shared identities and (5) interpersonal relationships (Baym 2010; Wellman et al. 2003). Previous research has expressed expectations of how virtual communities could help civil society function more effectively through the development and diffusion of online social networks (Jennings and Zeitner 2003; Livingstone and Markham 2008; Shah et al. 2005, 2001). Due to its increase in acceptance, many organizations and municipalities in Japan subsequently initiated projects to consider or implement Facebook usage for better civic engagement. This chapter focuses on one fascinating case in Ibaraki prefecture—specifically, the Tsukuba Civic Activities Cyber-Square Experiment [*Tsukuba Shimin Katsudō no Hiroba*]—of the Tsukuba municipal government of Japan.

The Tsukuba Civic Activities Cyber-Square Experiment

The Tsukuba municipal government of Ibaraki Prefecture in Japan is located approximately 60 kilometers northeast of Tokyo, the capital of Japan. The population is slightly more than 210,000. This municipality has been working jointly with the University of Tsukuba and Intel Corporation to create the Tsukuba Civic Activities Cyber-Square with the self-defined goals of: (1) nurturing future human resources and cultivating entrepreneurship; and (2) reactivating communities and recreating a healthy civic life by the year 2015. With these objectives in mind, in 2012, the Tsukuba municipal government began an empirical experiment to promote cooperation among citizens by creating a new perspective among citizens through social media. To facilitate this, the municipality set up a "cyber plaza" or "cyber-square" in Facebook to help promote networking among civic activities and groups. This cyber-square was initiated to create a foundation for information sharing and help visualize civic activities for enhancing civil society within the city. Prior to commencing this experiment, Tsukuba city already had the highest number of Facebook users in Ibaraki with approximately 10,860 users in October 2011, a figure calculated through access to the Facebook statistics that were available through the advertising interface that provides the number of users that have indicated Tsukuba as their domicile in their profile. With these intentions, this initial experimental period was initiated in January 2012 and ended its first experimental phase at the end of June 2012. With its success, social media usage was adopted and integrated into the city's operations in cooperation with the University of Tsukuba.

In this section, we report on this preliminary phase of the experiment by first comparing the types of activities featured on this Facebook page as well as other Facebook pages operated by municipalities or groups within Ibaraki prefecture. Then, we quantitatively examine the posts in the Tsukuba Civic Activities Cyber-Square from January 20 to June 30, 2012 that refer to social capital factors using SPSS computer-assisted text analysis. During that period, Tsukuba city experienced a local disaster on May 6, 2012 in the form of a rare tornado storm. Thus, within our analysis of the city's experimental period with Facebook, we also had the opportunity to assess firsthand how community information provision and communications through Facebook played a vital role during the emergency situation from May 6 to May 31, 2012.

Comparison with Other Social Networking Services in Ibaraki

In order to situate the city of Tsukuba's Facebook page among other locally operated Facebook pages in Ibaraki prefecture, we identified the top five Facebook sites in the same prefecture at the time of the first phase of the experimental period from January 20 to June 30, 2012 (see Table 4.1). Our selection criteria included sites that were either being operated by prefectural-level or local governments, or Facebook pages that had been in existence for a similar or longer duration than the city of Tsukuba's Facebook page.

Table 4.1 Top five Facebook sites originating in Ibaraki prefecture (January to June 2012)

Page title (English)	*Page title (Japanese)*	*Operating entity*	*Likes as of June 30, 2012*
Tsukuba Civic Activities Cyber-Square	*Tsukuba Shimin Katsudō no Hiroba*	Tsukuba city	1263
Foodies in Ibaraki	*Umaimono dokoro Ibaraki*	Ibaraki prefecture, Agricultural and fisheries distribution department	907
Enchantment of Ibaraki	*Ibaraki no miryoku-o tsutatetai*	Ibaraki prefecture, Public relations department	863
Mito City Tourism	*Mito-shi kankōka Mitoshikankōkyōkai*	Mito city, Tourism board	286
Sakuragawa City	*Sakuragawa-shi*	Sakuragawa city	491

The majority of the sites were focused either on promoting a particular feature of Ibaraki prefecture or tourism within the prefecture. For example, the "Foodies in Ibaraki" page concentrates on promoting local delicacies, as well as popular restaurants and eateries. "The Enchantment of Ibaraki" page focuses on encouraging tourism in Ibaraki by including information on local events, seasonal topics, popular tourist spots, and accommodation. The "Mito City Tourism" page is also about tourism, providing local information about the city of Mito (the capital of Ibaraki prefecture), its history, and its traditions. The "Sakuragawa City" page is a Facebook page that is more general in terms of content, offering information and topics about Sakuragawa city, tourist information about the area, as well as listing restaurant information and popular places for families to visit. Among the top five sites, the Tsukuba Civic Activities Cyber-Square is the only page that focuses solely on civic activities and civil society. Due to the lack of Facebook pages in Japan with similar focus on such activities at the time of the experiment, the four tourism-oriented Facebook pages were selected for comparison.

Figure 4.1 shows a comparison of the total number of posts on the five Facebook pages during the period from January to June 2012 when the Tsukuba Civic Activities Cyber-Square was in its initial experimental phase.

As noted earlier, providing tourism-related information, which is essentially oriented towards information provision rather than communication, was the main focus of four of the five Facebook pages. In our assessment

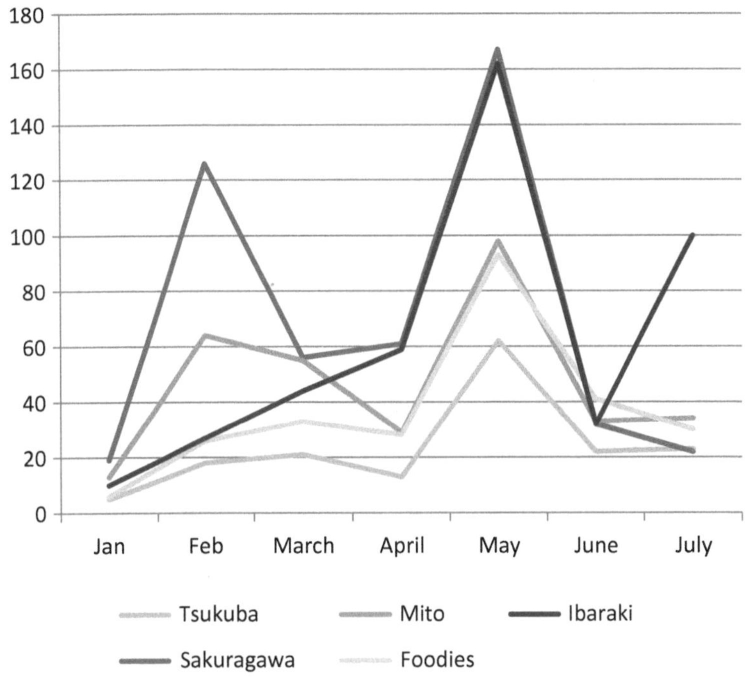

Fig. 4.1 Total posts on the top five Facebook pages in Ibaraki, January to June 2012

of the Facebook sites, we discovered that the sites that provided information regarding tourism or food (information on four out of the five Facebook sites that could be directed at a broad range of people including residents or visitors) were likely to have more posted comments and likes in comparison to information aimed at residents (such as that posted on the Tsukuba city page).

For example, in terms of tourism-related information, the increases in the number of posts in the Sakuragawa and Mito Facebook pages during February 2012 can be attributed to drawing potential visitors to local events such as the Makabe *Hina matsuri* (Makabe Doll Festival), a traditional local event that celebrates the Japanese custom of "Girl's Day" in early February, and the Sakuragawa annual marathon. Similarly, in Mito city during February 2012 and into early March, the majority of postings focused on the Mito city plum blossom festival, an eight-day festival that was held on four consecutive weekends from February 18 to March 11, 2012, and featured various theme events related to plum blossoms. The festival celebrates Mito city's official flower that figures prominently in the city's *Kairakuen* municipal park, known for being one of the three most beautiful municipal parks in Japan. All four tourism-related sites showed spikes in activity during the period late April to early May, coinciding with the "Golden Week" holiday period (wherein there are four national holidays occurring within the period April 29 to May 5), a popular period for domestic travel. Among the five Facebook sites, Fig. 4.1 also shows that the Tsukuba Civic Activities Cyber-Square had a competitive amount of posted comments and likes in comparison to the other four pages in consideration of the characteristics of information being communicated.

In terms of information provision, Facebook pages can be used in a similar manner as websites wherein the owner of an open Facebook page can transmit information to a broad audience. However, the engagement opportunities that are built into Facebook's architecture, such as the liking, commenting, and sharing features for viewers or users to interact with the owner of the page in real time, offer enhanced communications. Thus, we also assessed Facebook's engagement functions for the five Facebook pages (Fig. 4.2).

As illustrated in Fig. 4.2, a sudden increase can be observed during May 2012 in the Tsukuba Civic Activities Cyber-Square. Although the number of posts by the four other Facebook pages are greater than the Tsukuba Civic Activities Cyber-Square from January through April, access to engagement features surpasses the Sakuragawa page for May and June as well as the Mito page. The increase in online engagement in the Tsukuba Civic Activities Cyber-Square after May 6, 2012 is due to the tornado disaster that occurred around 13:00 that day and resulted in outages affecting 20,000 residences and deaths and injuries, along with destruction of homes and other facilities (this disaster will be examined in detail later in this chapter).

From this analysis, we observed that among the pages we selected, (1) social networking service pages that communicate tourism-related or recreational information are more likely to have comments posted or be liked, however, (2) social networking service pages that focus solely on communicating civic, resident-oriented, and government-provided information may be considered effective for increasing spontaneous feedback, such as comments or likes on Facebook.

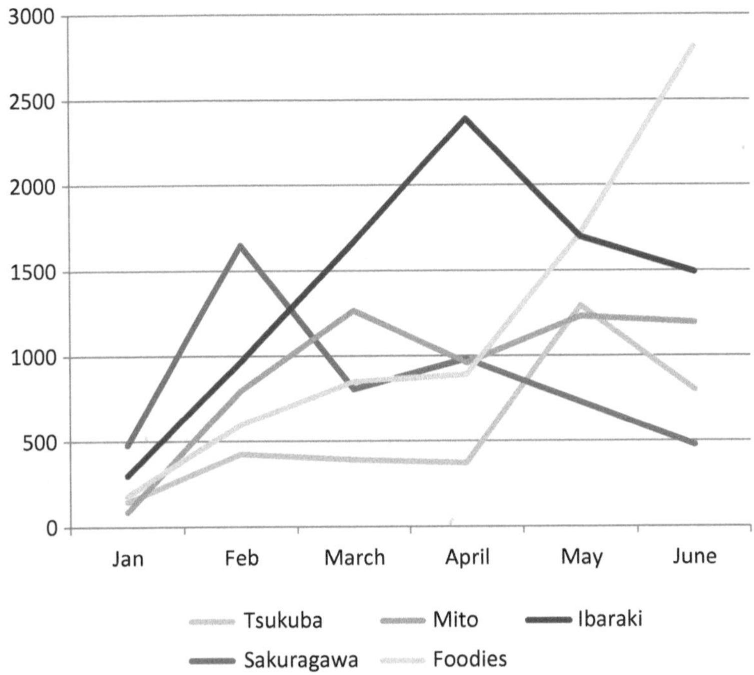

Fig. 4.2 Engagement features (total of likes, comments, and shares) on the top five Facebook pages in Ibaraki, January to June 2012

Analysis of Posts on the "Tsukuba Civic Activities Cyber-Square"

The Tsukuba Civic Activities Cyber-Square Facebook page had a total of 109 posts (inclusive of updates and also comments to updates) between January 20 and June 30, 2012. We divided the overall contents into two broad categories: The first category concerns posts related to everyday life (62 posts) and the second category is made up of content posted in reaction to the May tornado (47 posts). Computer-assisted text analysis has been used in studies to analyze sentiment in the past (Park et al. 2011); therefore, we used SPSS Text Analytics for Surveys 4.0 to analyze all postings morphologically after extracting the data. In this section, we first assess the posts related to everyday life in Tsukuba to find communality

Table 4.2 Analysis of posts in Tsukuba Civic Activities Cyber-Square

Morpheme	*Total posts (N) and %*
Gratitude	27 (43.5%)
Requests	16 (25.8%)
Affirmation	8 (12.9%)
Delight	6 (9.7%)
Encouragement	5 (8.1%)
Total	62 (100.0%)

with concepts related to social capital, such as community, public interest, social participation, and civic participation.

Computer-assisted text analysis allowed us to assess the 62 posts regarding everyday life that were made to the Tsukuba Civic Activities "Cyber-Square" during this period in terms of morphemes that reflect features of social capital. Table 4.2 shows the overall results of the morphemes that were extracted from the posts.

Posts that reflected gratitude (27 posts, 43.5%) and requests (16 posts, 25.8%) made up more than two-thirds of the total number of posts. Among the examples that we extracted, expressions of gratitude include phrases such as "Volunteering like this is welcome" and "Thanking you always." These two types of postings, comprising close to 70% of the total number of posts in the "everyday life" category, were deemed to reflect social support and reciprocation among the users of the Tsukuba Civic Activities Cyber-Square. The third type of posting, which we termed "affirmation," accounted for more than 10% of the posts, and we considered these types of posts to reflect a positive image towards the Cyber-Square Facebook page. Positive feelings regarding the local area (Harper 2002) were demonstrated by the six posts (9.7%) that expressed "delight." Finally, the social capital features of reciprocity and trust were illustrated by the five posts (8.1%) that were categorized as "encouragement."

Social Capital Through Updates

The results of the comparison of the activities within the Facebook page of the Tsukuba Civic Activities Cyber-Square and the other Facebook pages being operated by municipalities in the vicinity show that the total number of posts are lower overall in the Tsukuba Civic Activities Cyber-Square than on the other pages with "recreational" content. However, as described earlier, the Tsukuba Civic Activities Cyber-Square has been proven to be effective in nurturing social capital through a computer-assisted text analysis of the posts made in the Tsukuba Civic Activities Cyber-Square. Even without making the content artificially "recreational," communication to enhance civil society through social networking services has been demonstrated to be a viable option. Through the example shown earlier, the Tsukuba Civic Activities Cyber-Square Facebook page can gain attention equal to the other Facebook pages of local municipalities in the same region.

Assessing posts during the month after the natural disaster in the Tsukuba area demonstrates that civic-related information such as disaster relief communication and civic activity notices are equally valuable for users of SNSs with such orientation, and can be competitive when comparing the number of likes or comments. Municipalities need not post "recreational" or tourism-oriented information continuously to elicit instantaneous reactions such as likes or potentially superficial comments to their updates. People who have or who developed an interest in the communications of the municipality will "like" the page and subscribe to the disseminated information. It is possible that when first accessing the page, a user might be a "silent subscriber," or someone who reads posts but does not react or comment on them. This may continue until a time comes when the subscriber feels the need to comment or communicate in another way. When a person is motivated to communicate and participate in communications involving the municipality, the person becomes more involved in the community, in this case, through civic activities. This interest can resonate from cyberspace back into real society. This cycle of interest and involvement demonstrates the possibility of SNS usage among the local municipalities in Japan. We shall now examine the flow of "civic" information of disaster relief communication during the Tsukuba Tornado disaster in further detail.

The May 6, 2012, Tsukuba Tornado (Tsukuba, Ibaraki)

In comparison to other regions of the world, while not unheard of, tornados in Japan are relatively rare, according to the Japan Meteorological Agency. Tornado sightings in Japan are not regionally isolated, and therefore, they may occur anywhere throughout the country. In the five-year period spanning 2005–2011, 116 tornados were recorded (on average,

23.2 per year, not including tornados on the ocean) by the Japan Meteorological Agency. Tornados occur most often in September and October, known as the typhoon season in Japan, and are relatively rare from January till May in comparison.

On May 6, 2012, at approximately 13:00, the southwest area of Ibaraki prefecture (where the city of Tsukuba is located) and southeast area of Tochigi prefecture were struck by a powerful tornado. The storm took a northeast route, first heading into the Hojō area in the north part of Tsukuba and then continuing in a northerly direction to neighboring Tochigi prefecture. Factories and buildings located in Tsukuba city's northern industrial area as well as the Hojō area were destroyed or had their roofs blown away. Ultimately, approximately 2000 buildings and homes were damaged, with Tsukuba city being the hardest-hit area in terms of overall damage. According to the Ministry of Internal Affairs and Communications of Japan, over 800 houses were damaged and 170 were completely destroyed. One person was buried alive in a house collapse and subsequently died. Immediately after the tornado, power outages affected 21,000 homes and continued in the area for several days. Damage to water supply systems meant that close to 5200 homes also temporarily lost water access.

The Japan Meteorological Agency observed so-called "supercell" thunderclouds due to a cold air mass over Japan developing at the time of the tornado occurrence (*The Daily Yomiuri* 2012). According to the Fujita-Pearson Tornado Scale, or F-Scale (a six-level scale ranging from F0 being the weakest to F5 as the strongest), the tornado in Tsukuba was categorized as a F2 tornado with winds of more than 50 meters per second or 180 kilometers per hour, and may have been temporarily as strong as an F3 in some areas (*The Daily Yomiuri* 2012). The most powerful tornados ever recorded in Japan have been F3 tornados, such as the November 7, 2006 tornado in Hokkaido, the September 24, 1999 tornado in Aichi Prefecture and the December 11, 1990 tornado in Chiba Prefecture, which tore down houses and upended train cars.

Tsukuba Civic Activities Cyber-Square and the Tornado

The Tsukuba municipal government went into action immediately after the initial sighting and confirmation of the tornado hitting the northern area of Tsukuba city. The city government very rapidly created a disaster response center in its main Municipal Hall. Through these circumstances, the Tsukuba Civic Activities Cyber-Square Facebook page immediately took an emergency communications channel role during its experimental phase from May 6 to May 26, 2012, as it evolved into one of the essential information tools of the Tsukuba municipal government for sending out and collecting information pertaining to the tornado.

Figure 4.3 shows the number of postings during this three-week period. A posting by the municipality notifying residents that a tornado had hit the northern area of Tsukuba on May 6, 2012 had a reach of over 1000 with subsequent postings of the damage of the area. A subsequent posting on May 6, 2012 included a notification that an operations center had been established at the *Tsukubane* citizens' hall in northern Tsukuba with first-aid facilities. On May 7, the day after the tornado, volunteers were being recruited along with postings through Facebook of information about volunteer and clean-up activities. The procedures for accepting waste materials from the disaster were also being communicated via Facebook. On May 8 and 9, notices regarding volunteer activities and relief materials were also posted. On May 10, more detailed information related to the May 6 tornado started being reported through the Facebook page along with a new weather alert of another possible tornado that same day.

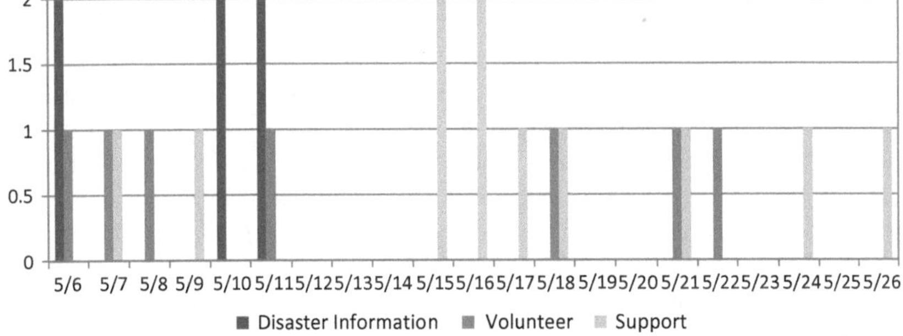

Fig. 4.3 Dates of postings and categories from May 6 to May 26, 2012, Tsukuba Civic Activities Cyber-Square

On May 11, the Tsukuba Civic Activities Cyber-Square made a posting about availability of the Hojō *Machikado* newspaper, printed to allow residents without Internet access to obtain information. (The Hojō *Machikado* newspaper is published by the local area promotion board.) The Facebook page also informed citizens of extended working hours of municipal consumer support centers and sent out warnings about uniformed swindlers impersonating municipal workers, trying to cheat or extort money out of disaster victim residents for waste disposal services.

After the tornado volunteer center closed on May 14, postings on May 15 began focusing on health consultation for disaster victims and information on application procedures for official disaster certificates for tax exemptions, with other how-to information on official procedures being communicated through various channels. On May 17 and 18, news about disaster victim benefits such as free transportation, as well as briefings on reconstruction and support centers for residents, were publicized through the Tsukuba Civic Activities Cyber-Square. On May 21, information related to subsidies and loans were posted and the Facebook page for the Hojō *Machikado* newspaper was introduced. Live benefit concert information was posted on May 22, followed by further information on waste disposal two days later on May 24.

All postings from May 6 through May 16 had a reach of over 700, and subsequent postings had similar numbers. The May 26 posting related to the tornado relief concert and charity flea market for disaster-affected mothers and children had the lowest reach of 430, although close to 7% of the viewers reposted the information (Table 4.3).

As can be observed from Table 4.3 and Fig. 4.3, the postings can possibly be divided into three categories: (1) disaster information, (2) volunteering and (3) support. Initial information posted on the Tsukuba Civic Activities Cyber-Square was clearly about disaster information, however one week after the tornado, the main topic of the postings turned to support information for the disaster victims. Information related to volunteering was periodically distributed through the postings and actively assisted relief efforts. Information was disseminated swiftly during this period, although interactivity was not as high as one should expect from the mechanism of social networking sites such as Facebook.

The number of likes was higher than the preceding period of the Tsukuba Civic Activities Cyber-Square experiment, and the number of responses shows that information was being disseminated effectively. As the topic turned to support for the disaster victims one week after the

Table 4.3 Information about posts during tornado disaster: Tsukuba Civic Activities Cyber-Square

Date	*Category*	*Reach*	*Action by user*	*Talking about*	*Communicated*
2012/5/6	Disaster Info	1023	153	105	10.3%
2012/5/6	Disaster Info	847	68	48	5.7%
2012/5/6	Volunteer	771	60	40	5.2%
2012/5/7	Volunteer	882	111	73	8.3%
2012/5/7	Support	768	67	41	5.3%
2012/5/8	Volunteer	823	75	47	5.7%
2012/5/9	Support	780	66	44	5.6%
2012/5/10	Disaster Info	749	44	31	4.1%
2012/5/10	Disaster Info	740	62	31	4.2%
2012/5/10	Disaster Info	723	28	21	2.9%
2012/5/11	Volunteer	875	83	57	6.5%
2012/5/11	Support	776	36	29	3.7%
2012/5/11	Support	770	53	42	5.5%
2012/5/15	Support	797	39	31	3.9%
2012/5/15	Support	733	17	14	1.9%
2012/5/16	Volunteer	762	42	31	4.1%
2012/5/16	Volunteer	547	36	3	0.6%
2012/5/17	Support	702	21	14	2.0%
2012/5/18	Support	727	44	31	4.3%
2012/5/18	Volunteer	700	35	28	4.0%
2012/5/21	Support	673	22	17	2.5%
2012/5/21	Volunteer	663	53	33	5.0%
2012/5/22	Volunteer	720	51	31	4.3%
2012/5/24	Support	631	21	15	2.4%
2012/5/26	Volunteer	430	46	30	7.0%

tornado, the number of postings regarding information or facts about the tornado decreased in comparison to the first week immediately after the tornado (Fig. 4.4).

Tsukuba Civic Activities Cyber-Square Experimental Period

This chapter analyzes the initial experimental phase of SNS usage by the Tsukuba municipal government in Japan, focusing on the possibilities and problems of complementary communication channels. We described how municipalities in Japan are using SNS to enhance civil society and how these services can provide vital information and connect citizens, municipal

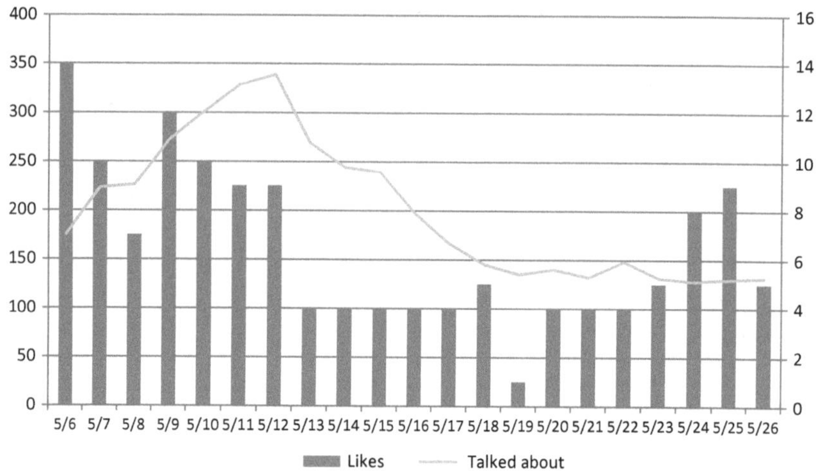

Fig. 4.4 Likes and talked about concerning the May 6, 2012, tornado disaster, Tsukuba Civic Activities Cyber-Square (May 6 to May 26, 2012)

governments, and civil society. Although we speculated on the merits and demerits of the use of social networking services by municipal governments, our analysis of the practical experiment conducted by the Tsukuba municipal government during this six-month period highlighted both positive and negative aspects.

The main obstacles to initiating SNS usage by the Tsukuba municipal government prior to the experiment stemmed from common anxieties held by many local governments. The strongest of these fears was the risk of *Enjyo* or being crucified by the Japanese Internet users (See Chap. 1). The fears of making mistakes in the postings and/or inadvertently spreading inappropriate information through social media, identified as common concerns among Japanese municipalities in the *White Paper* issued by the Ministry of Internal Affairs and Communications (2013), were also shared by administrators within the Tsukuba municipal government. These past negative examples or instances in Japan have resulted in delaying or prohibiting social media usage among many organizations and have created a general fear of using social media among those with less developed skills in managing or utilizing information technology. However, instead of shelving plans to use social media due to fear of trolls, the creation of guidelines was a vital step in the process of implementing the Tsukuba Civic Activities

Cyber-Square to avoid unwanted consequences of Internet usage that have been discussed in previous studies (Christopherson 2007; Dwyer et al. 2007; Tufekci 2007).

The Tsukuba guidelines required that the municipality utilize social media that allowed users to use their real names, created procedures to avoid careless mistakes, had workers refrain from posting slanderous information, leaking secret and/or private information about the organization, avoid posting uncertain facts and clearly indicate when posting opinions that are the official position of the government. These guidelines were established through discussions within the Tsukuba government prior to the experiment. Many features of Facebook coincidentally satisfied the guidelines of using real names (in contrast to the Japanese SNS *Mixi* which can be used anonymously) and were vital elements for initiating Facebook usage and completing this experiment to a successful first stage.

The Tsukuba Civic Activities Cyber-Square built up a following of people who are relatively interested in civic activities and who can be mobilized quickly to disseminate information about volunteering, especially in the case of disaster relief as seen during the May 2012 tornado. SNS can therefore play a vital role in filling in the gap between the ordinary citizen and government, allow for more transparency, and create a community in cyberspace with people who have common interests. As we know from sociological studies, people within a similar environment with similar characteristics and similar experiences tend to form communities (homophily). Homophily within Facebook, especially with similar orientations (in this case, towards government and civic activities) is natural behavior, and with more time and accompanying the increase of users, the Facebook community can reflect onto the real world and society what is currently being seen in cyberspace. Outside of the context of this example, this may be seen as a problematic characteristic of SNS and cyberspace, however, in the case of government and civic activities, this can be viewed as a possibility.

In this chapter, we also examined how social media began to play a prominent role in information provision and communications during and after a disaster. The increase in Facebook usage in Japan has seen numerous municipalities attempting to use this platform to communicate more effectively with local citizens in attempts to cope with disasters. The example of Takeo city in Saga is one example, along with many others beginning to acquire Facebook accounts and attempting to reach out to local residents through this platform. The Tsukuba Civic Activities Cyber-Square

Experiment is another example of how municipalities are endeavoring to use SNS platforms such as Facebook to improve communications and create positively oriented cyber communities in Japan. The findings here are confined to activities involving one municipality in Japan and one natural disaster, however, our results can be thought to be applicable in many situations.

The tornado which hit Tsukuba city of Ibaraki prefecture in Japan on May 6, 2012 tested the municipal government's ability to handle a disaster, but also was an opportunity to observe how people would communicate on Facebook during a crisis through the Tsukuba Civic Activities Cyber-Square. The reach and reactions towards the posts on the Tsukuba Civic Activities Cyber-Square displayed considerable consistency with the previous literature advocating the potential effectiveness of social media during emergencies and disasters. We have observed that through social media, a communication network of social support was formed in Tsukuba when experiencing natural disasters like these, and social interaction was made possible in a wide region through social networking services.

During this tornado-inflicted disaster, the Tsukuba municipal government initially sent out disaster-related information and, quickly after, began sending out information about disaster volunteers and information on procedures for volunteering. In the days following the disaster, postings shifted towards more support for the disaster victims and information on events and ways to help these people. The occurrence of this tornado disaster in Tsukuba coincided with the Tsukuba Civic Activities Cyber-Square Experiment, so although the natural path of diffusion among the residents of Tsukuba city was altered, it was also an opportunity to demonstrate how social media such as SNS can play a possibly effective role in disaster-related information notification, enhancing civil society through volunteer coordination and later support and relief efforts.

Since the end of the experimental phase, the Tsukuba Civic Activities Cyber-Square has been incorporated into the daily operations of the Tsukuba municipal government. An increase in the size of the SNS community is fostering future possibilities of civic activities in the Tsukuba city area and offers other ideas and solutions to local municipalities on how to deploy SNS to further their causes. Many local governments in Japan are initiating the usage of social media platforms to virtually connect community members to confront an increasing number of social issues. These communities are following the steps taken by the Tsukuba municipal government to redesign themselves through using social media and recreate their communities.

References

Appadurai, A. (1995). The production of locality. In R. Fardon (Ed.), *Counterworks. Managing the diversity of knowledge*. London: Routledge.

Baym, N. K. (2010). *Personal connections in the digital age*. Cambridge: Polity.

Carey, J. (1989). *Communication as culture: Essays on media and society*. New York: Routledge.

Christopherson, K. M. (2007). The positive and negative implications of anonymity in internet social interactions: On the internet, nobody knows you're a dog. *Computers in Human Behavior, 23*(6), 3038–3056.

Dwyer, C., Hiltz, S. R., & Passerini, K. (2007). Trust and privacy concern within social networking sites: A comparison of Facebook and MySpace. *Proceedings of AMCIS 2007*.

Harper, R. (2002). *The measurement of social capital in the United Kingdom*. Country Paper prepared for OECD-ONS International Conference on Social Capital Measurement, London.

Hashimoto, Y., & Ohama A. (2014). The role of social media in emergency response: The case of the Great East Japan Earthquake. *NIDS Journal of Defense and Security, 15* (2014). Retrieved August 7, 2015, from http://www.nids.go.jp/english/publication/kiyo/pdf/2014/bulletin_e2014_6.pdf

Hooper, M. K. (1999). Disaster preparedness: An analysis of public safety agency and community preparedness during the Northridge earthquake. *International Journal of Public Administration, 22*(5), 679.

Howard, P. N., & Parks, M. R. (2012). Social media and political change: Capacity, constraint, and consequence. *Journal of Communication, 62*(2), 359–362.

Howard, P. N., Duffy, A., Freelon, D., Hussain, M., Mari, W., & Mazaid, M. (2011). *Opening closed regimes: What was the role of social media during the Arab spring project on information technology and political Islam*. Seattle: Department of Communication, University of Washington.

Hughes, A. L., & Palen, L. (2010). Twitter adoption and use in mass convergence and emergency events. *International Journal of Emergency Management, 6*(3–4), 248–260.

Inoue, T. (2013). Toward the operation guidelines for municipal social media account: From the analysis of user's behavior on twitter in the Great East Japan earthquake. *International Journal of Information Society, 59*(3), 137–145.

Jennings, M. K., & Zeitner, V. (2003). Internet use and civic engagement – A longitudinal analysis. *Public Opinion Quarterly, 67*(3), 311–334.

Kakei, Y. (2013). *Sōsharu Desain Jissen Gaido [A practical guide to social design]*. Tokyo: Eiji Press.

Kazuma, Y. (1995). On-line services to the rescue. *Japan Echo, 22*(2), 28.

Kroll-Smith, S. (2001). The Northridge earthquake: Vulnerability and disaster. *Contemporary Sociology, 30*(2), 174.

Livingstone, S., & Markham, T. (2008). The contribution of media consumption to civic participation. *British Journal of Sociology, 59*(2), 351–371.

McAtee, A., & Wolak, J. (2011). Why people decide to participate in state politics. *Political Research Quarterly, 64*(1), 45–58. doi:10.1177/1065912909343581.

Mills, A., Chen, R., Lee, J. K., & Rao, H. R. (2009). Web 2.0 emergency applications: How useful can twitter be in emergency response? *Journal of Information Privacy & Security (JIPS), 5*(3), 3–26.

Ministry of Internal Affairs and Communications, Japan (MIC). (2013). *Heisei Nijugonendo Jōhō Tsushin Hakusho. [Information and Communications White Paper 2013].* Tokyo: Ministry of Internal Affairs and Communications. Retrieved August 11, 2015, from http://www.soumu.go.jp/johotsusintokei/whitepaper/h25.html

Noam, E. M., & Sato, H. (1995). Kobe's lesson: Dial 711 for 'open' emergency communications. *Telecommunications Policy, 19*(8), 595.

Park, S. J., Lim, Y. S., Sams, S., Nam, S. M., & Park, H. W. (2011). Networked politics on Cyworld: The text and sentiment of Korean political profiles. *Social Science Computer Review, 29*, 288–299.

Pedersen, M. (2003). *Between homes: Post-war return, emplacement and the negotiation of belonging in Lebanon* (Working Paper No. 79). UNHCR, The UN Refugee Agency. Retrieved July 23, 2008 from http://www.unhcr.org/3e59e16e4.html

Putnam, R. D. (2000). *Bowling alone: The collapse and revival of American community.* New York: Simon & Schuster.

Resnick, P., Konstan, J., Chen, Y., & Kraut, R. E. (2012). Starting new online communities (Chap. 6). In R. E. Kraut & P. Resnick (Eds.), *Building successful online communities: Evidence-based social design.* Cambridge, MA: MIT Press.

Rheingold, H. (1993). *Homesteading on the electronic frontier.* Reading: Addison-Wesley.

Shah, D. V., McLeod, J. M., & Yoon, S. H. (2001). Communication, context, and community – An exploration of print, broadcast, and internet influences. *Communication Research, 28*(4), 464–506.

Shah, D., Cho, J., Eveland, W., & Kwak, N. (2005). Information and expression in a digital age – Modeling internet effects on civic participation. *Communication Research, 32*(5), 531–565.

Sutton, J., Palen, L., & Shklovski, I. (2008, May). Backchannels on the front lines: Emergent uses of social media in the 2007 Southern California Wildfires. In F. Fiedrich & B. Van De Walle (Eds.), *Proceedings of the 5th international ISCRAM conference* (pp. 624–632). Washington, DC.

The Daily Yomiuri (2012, May 8). Temperature differences spawned deadly tornado. Daily Yomiuri Online. Retrieved September 1, 2012 from http://www.yomiuri.co.jp/dy/national/T120507005103.htm

Tufekci, Z. (2007). On the Internet, everybody knows you're a dog: Presentation of self for everyday surveillance. *Paper presented at the American Sociological Association, 2007.*

Wellman, B., & Haythornthwaite, C. A. (Eds.). (2002). *The internet in everyday life*. Malden: Blackwell Pub.

Wellman, B., Quan-Haase, A., Boase, J., Chen, W., Hampton, K., Isla de Diaz, I., & Miyata, K. (2003). The social affordances of the internet for networked individualism. *Journal of Computer-Mediated Communication, 8*(3), 0.

White, C. (2010). Social media and meta-networks for crisis mapping: Collaboratively building spatial data for situation awareness in disaster response and recovery management. Paper presented at *2010 Specialist Meeting Spatio-Temporal Constraints on Social Networks*. University of California, Santa Barbara, Center for Spatial Studies, December 14–15, 2010. Retrieved August 11, 2015, from http://www.ncgia.ucsb.edu/projects/spatio-temporal/docs/White-position.pdf

Muneo Kaigo is an associate professor in the Faculty of Humanities and Social Sciences at the University of Tsukuba in Japan. He teaches courses in media communication in the Graduate School of Humanities and Social Sciences, media management in the Graduate School of Business Sciences and communication science in the College of Comparative Culture at the University of Tsukuba. He is currently leading a joint research project on social media uses among local municipalities in Japan with cooperation from the municipal government of Tsukuba and Intel Corporation Japan.

Leslie Tkach-Kawasaki is an associate professor in the Faculty of Humanities and Social Sciences at the University of Tsukuba in Japan. Her research interests are mainly centered on how Japanese political actors utilize diverse Internet-based media channels for electoral campaigning and communication with the electorate. Recently, she has been involved in expanded projects involving the history of the Internet in Japan, Japanese politicians and social media network analysis, as well as cross-cultural differences in ethical and methodological approaches to Internet-based research.

CHAPTER 5

Promotion and Care of Online Communities: Necessary Elements for a Self-Sustainable Online Facebook Community

Muneo Kaigo and Sae Okura

Challenges in SNS Usage by Japanese Local Governments

Collaboration between the public sector and the private sector is one of the most important areas of current research into public policy studies. Nishio (2004) defined collaboration as the process of multiple entities such as individuals as well as groups that have different abilities and roles that complement each other. They all work together continuously on even ground toward the achievement of a common goal. The concept has been discussed in the context of participatory democracy (Kodagiri 2014). Theories in this area often focus on the educational effects of political participation and argue that opportunities to participate in the policy process can promote citizens' democratic values. As a result, participatory democracy has the potential to stabilize the political system (Kabashima 1988), meaning

M. Kaigo (✉)
Faculty of Humanities and Social Sciences, University of Tsukuba,
Tsukuba, Ibaraki, Japan

S. Okura
Faculty of Humanities, Law and Economics, Mie University,
Tsu, Mie, Japan

M. Kaigo (ed.), *Social Media and Civil Society in Japan*,
DOI 10.1007/978-981-10-5095-4_5

that citizens become more democratic when given the proper opportunities to participate in politics through a collaborative environment.

In the past, we have seen how collaboration between the public sector and the private sector is facilitated through social media. Social media not only enables open government (Lee and Kwak 2012), but also provides platforms allowing citizens better access to public services and to become more involved in policy-making (Linders 2012). In addition, the public sphere expands when social media brings governments and citizens together (Wilhelm 2000; Sassi 2000; Keane 2000; Dahlberg 2001). However, some scholars disagree with this argument. For example, Poster (1995) pointed out how Internet users can change their names easily, so the Internet as a new communications technology would have a limited effect with regard to changing the political apathy of citizens. For studies focusing on Japan, Tanaka and Yamaguchi (2016) have highlighted similar arguments. However, such concerns have already been addressed, at least partially, through the introduction and enforcement of real-name user registration by social media platforms such as Facebook. Social media can disseminate information and knowledge that may be of use to citizens. In addition, we have already reviewed how social media can play a useful role in disaster prevention (Kavanaugh et al. 2012). As we have discussed in Chap. 2, in particular, it enables governments to continue to function and to communicate when large-scale disasters prohibit normal administrative activities and communications. In other words, social media makes it easier for the public sector and the private sector to interact, thereby allowing more people to join the political discourse and enabling collaboration between the two sectors.

This chapter aims to identify the various difficulties faced by employees of local governments when they try to make use of social networking services (SNSs) to share information. The authors conducted exploratory interviews with three employees of the Tsukuba municipal government who had been involved in the operation of the municipal government's official Facebook page—the Tsukuba Civic Activities Cyber-Square. Ultimately, the goal is to create a list of challenges in SNS use that are faced by local governments in Japan as this may provide hints for other nations as well. As this chapter is a pioneering work in relation to such a subject and is based only on a small sample, further research is necessary before we generalize our findings. Nevertheless, we believe that this chapter's results will be useful in understanding the evolving phenomena of local government use of SNSs to disseminate information and participate in civic activities. The first section of this chapter will review previous work on related subjects, then explain our objectives and methodology, and finally provide our interview results and discuss our findings before we proceed to the second section of this chapter.

Resources and Skills for Operation

Most studies on SNSs focus on whether using SNSs has led to desired outcomes. In Japan-related case studies, for example, Shoji (2008) found that strong ties were formed among local SNS users who communicated actively with one another. Hence, social network services can help: (1) make everyday life more convenient; (2) organize events and community-building activities; (3) attract more consumers to the shopping street and boost sales; and (4) bring more tourists. Eventually, such merits will lead to regional revitalization. Noguchi and Ito (2013) studied the city of Takeo in Saga Prefecture, where the local government pioneered the use of Facebook for its public relations and other means of both internal and external communication. They found that Facebook not only created more social capital among government employees but also successfully improved local "civic power" (Sakamoto 2010) by strengthening the connection between citizens and the government (Noguchi and Ito 2013). In other words, studies on the effects of SNS usage have shown that these services can expand the public sphere when properly utilized, although the use of SNSs do not always lead to desired outcomes.

Some studies have focused on the efficiency of SNS usage, attempting to identify factors that inhibit the use of the services. Nakano (2014b) conducted semi-structured interviews with 14 local governments that used the Japanese custom *Chiiki SNS* platforms and studied how local government employees were involved in the management and operation of their *Chiiki SNS*. Although a few exceptions exist, he found that as long as the *Chiiki SNS* systems were functioning, local government employees kept their involvement to a minimum level because personnel and budget cuts meant that they were forced to manage the *Chiiki SNS* systems while holding other posts. Under such circumstances, it was difficult to achieve efficient *Chiiki SNS* use without having any dedicated staff managing the social media platform and related technology on a full-time basis, or having a dedicated section of the government (Nakano 2014b). Using results from the same interview, Nakano (2014a) concluded that three factors could lead to a local government deciding to terminate local *Chiiki SNS* use:

(1) project evaluation; (2) the introduction of a time limit; and (3) competition with other existing ICTs. Studies like these emphasize the importance of resources (e.g., staff and budget) devoted to the SNS projects.

Red Tape

Another area of study regarding SNSs examines the regulatory and bureaucratic factors related to the governmental body that runs these operations. In most nations and local governments, numerous rules, regulations, and procedures are in place to ensure that the organization can run and operate smoothly. Governments also pass various laws, ordinances, and regulations in order to achieve certain goals to meet objectives that have been set by lawmakers. However, excessive formalities can, over time, become administrative burdens, what is colloquially known as "red tape" (Kaufman 1977). Bozeman (1993) defined organizational red tape as "rules, regulations, and procedures that remain in force and entail a compliance burden for the organization but have no efficacy for the rules' functional object."

A fairly extensive amount of literature on red tape exists in the field of public administration studies (Merton 1949; Feeney 2011; Bozeman and Feeney 2011; Feeney 2012; Riccucci 2012; Borry 2013). For example, Bozeman and Feeney (2011) classified the delays in administrative activities into two types or categories: red tape that is "built-in" when the rules are first enacted, including bad or dysfunctional laws and regulations and complex procedures. This is referred to as "rule-inception red tape." The red tape occurring as rules, regulations, and procedures that are implemented through the evolution of rules and procedures is referred to as "rule-evolved red tape." Merton (1949) argued that the structures of bureaucracies have a tendency to move toward dysfunction and introduced the idea of "dysfunctions of bureaucracies." He also introduced and connected the use of the terms, "formalism," "ritualism," and "red tape" as examples of these types of dysfunctions.

Scholars argue that organizational red tape is one of the possible factors that impede the utilization of SNS by governments. When Shoji (2012) investigated electronic bulletin board system operations by local governments, out of the top 16 local governments that actively used electronic bulletin boards in 2003, only two had increased their use of the platform two years later, while six ended their use altogether and eight substantially reduced their usage. He concluded that personnel changes involving electronic bulletin board management staff were one of the major reasons for reduced activity. As previously mentioned in Chap. 3, Kogawa (2012) studied the development of Fujisawa Citizens' e-Conference Room, which was a joint operation conducted by the Fujisawa municipal government and Keio University. Between the platform's inception in 1997 and 2011, when the municipal government eventually withdrew from its direct management, Kogawa found that the amount of activity decreased after 2004 when the full-scale privatization of the platform failed to take place. Local government employees felt strong pressure to maintain the e-Conference Room as a centerpiece business of the city and were hesitant to carry out "irresponsible reforms." Cullen (2008) compared the cases of Japan and New Zealand and found that the Japanese public tended to be more skeptical of their government's ability to protect personal information and privacy. As a result, government employees in Japan have, in general, faced stricter monitoring and continue to have to deal with more restrictions.

Based on previous studies, the lack of full-time staff and IT skills along with the *nigate-ishiki* toward ICT that result in hesitation to better manage SNSs, as well as bureaucratic red tape can also make it difficult to utilize social networking services. However, there has been no comprehensive investigation of the challenges to SNS use faced by local governments, at least in Japan. Therefore, this chapter aims to generalize the difficulties faced by local governments when using Facebook by examining the case of the Facebook page "Tsukuba Civic Activities Cyber-Square."

The Organization and Management of the Tsukuba Civic Activities Cyber-Square

According to a March 2013 survey reported by the *Chiiki SNS Kenkyūkai* (*Chiiki SNS* study group), 466 Facebook pages were being managed by local Japanese governments. We cross-referenced the list of Facebook pages identified by the study group and confirmed that, as of July 2016, of the 466 pages on the list, 425 pages are still in existence and six pages deal with civic activities. Of these, the Tsukuba Civic Activities Cyber-Square had the most "fans" and the second-highest level of "engagement." This consistent level of activity explains why the Tsukuba Civic Activities Cyber-Square can be considered a representative case for examining civic engagement and civic activity revitalization via social media in Japan.

Within the Tsukuba municipal government, the Civic Collaboration Desk of the Civic Activities Division of the Civil Affairs Department is in charge of the daily operation of the Tsukuba Civic Activities Cyber-Space.

The Civil Affairs Department is composed of the Civil Affairs Division, the Civic Activities Division, the International and Cultural Affairs Division, the Sports Promotion Division, and the Lifelong Learning Division, and the Civic Activities Division is responsible for the protection of human rights, promotion and regulation of civic engagement, and communications with various self-government associations and neighborhood associations. Within the Civic Activities Division, the Civic Collaboration Desk is mainly in charge of the Tsukuba Civic Activities Cyber-Square page.

The core duties related to the operation of the Tsukuba Civic Activities Cyber-Square include: (1) management and operation of the Facebook page; and (2) organization of events related to the Facebook page. The first duty involves interviewing local citizens' groups, NPOs, and neighborhood organizations. Based on the interviews, reports are written and then regularly posted on the Facebook page. It also involves monitoring the posts made by the general public, filtering out any posts that are for solely commercial purposes and deleting any comments that may be considered improper. In addition, the Civic Activities Division is also responsible for replying to the various questions and comments that are posted on the page. The second duty involves organizing social gatherings for users of the Tsukuba Civic Activities Cyber-Square to communicate in real-life settings and promote information about the local festivals.

Interviews

We interviewed regular and contract employees of the Tsukuba municipal government who worked at the city hall and were involved in the operation of the Tsukuba Civic Activities Cyber-Square. Between the fiscal years of 2011 and 2014, a total of 17 local government employees were involved in the operation of the Facebook page (Table 5.1). These employees were divided into three groups: (1) those who were mainly involved in event organization; (2) those who were mainly involved in interviews and posting on the Facebook page; and (3) those who were involved in both of these activities. Of the 17, we interviewed three employees from the third group, and selected one from each fiscal year spanning from 2011 to 2014. For those who were involved in the project for more than two years, each full year of involvement was counted as "1."

We selected three employees (employees A, B and C) for our interviews. Employee A was from the Planning Division of the Planning

Table 5.1 Operation system (unit: person) of the Tsukuba Civic Activities Cyber-Square

	Events only	*Facebook operation only (interview and post)*	*Both*	*Total*
FY2011	–	4(2)	–	4(2)
FY2012	3(2)	0	1	4(2)
FY2013	2(2)	0	3	5(2)
FY2014	2(2)	0	2	4(2)
Total	7(6)	4(2)	6	17

1 The numbers in this table indicate the number of people at the end of each fiscal year

2 Numbers inside the parentheses indicate the number of people holding managerial posts

3 Each employee who remained engaged in the project throughout the fiscal year is counted "1" for the fiscal year

Department and was involved in the early stages of the tri-party (Tsukuba City, University of Tsukuba, Intel Corp.) joint project. Our purpose in selecting employee A to investigate the possible inhibiting factors in usage during the initial and early stages. Employee B became the staff manager of the page after the Tsukuba Civic Activities Cyber-Square entered its normal phase of operation. We selected employee B to investigate factors inhibiting elements of the daily operation of the page. Employee C was a contract employee hired to manage the Facebook page on a full-time basis. We considered employee C as a proper sample because the person was actively engaged in interviewing local citizens' groups and posting interview reports, as well as organizing and participating in running the events. The interviews were conducted in January and February 2015.

In an exploratory effort, we conducted semi-structured interviews with this small sample. The purpose of this study is to identify the difficulties faced by managers when utilizing social networking services in a governmental setting. In order to take into account as many factors that could affect SNS usage as possible, such as organizational structure and administrative efficiency at the city hall, we decided to conduct semi-structured interviews (For information related to factors promoting Facebook usage found in the same interviews, see Kaigo and Okura (2016)). In addition, because of the small number of existing cases studying local governmental use of Facebook, we believe it is proper to perform an exploratory analysis, which allows for more inclusive information collection through the interviews.

Emails explaining the purpose of the interviews were sent to the interviewees in advance. An interview was scheduled and conducted after the interviewee fully understood the purpose of the interview. All of the interviews were recorded using a digital voice recorder with permissions from the interviewees.

By conducting the interviews and analyzing our results, we categorized what Table 5.2 summarizes as the nine difficulties associated with project management identified throughout this section. The rest of this section reviews the evidence collected from the interviews. Statements made by the interviewees are emphasized using quotation marks, with the ID of the interviewee (A–C) inside parentheses.

i. Appreciative Assessment and Support from the Division in Charge

As previously mentioned, the Tsukuba Civic Activities Cyber-Square was launched as a tri-party project involving the Tsukuba municipal government, the University of Tsukuba, and the Intel Corporation. The municipal government's liaison with the other two parties is the Planning Division of the Planning Department, which proposed the Tsukuba Civic Activities Cyber-Square to the Civic Activities Division.

Of the three interviewees, employee A had the most knowledge about the details of this aspect of the project. According to employee A, who was involved in the launch of the project as a staff member of the Planning Division, when employee A was trying to hand the Planning Division's proposal to the Civic Activities Division, the key challenge was to make sure that the division in charge could have a full understanding of the purpose of the project. First, because Facebook was still "little known" inside the city hall at the time, it was necessary to persuade the director of the Civic Activities Division to create his own Facebook account and become familiar with this tool. Second, in order to institutionalize the project, it was necessary to request permission from the mayor to include the project in the municipal government's budget. Although the staff would prepare the budget request documents, it would be the director of the division in charge who would "bear the full brunt of" securing the budget. Thus, it would be difficult to move the project beyond test operation without the director's full comprehension and support (A).

Table 5.2 Analysis of interview results

	Selected comments by the interviewees
(1) Appreciation and support from the division in charge (A)	Besides the fact that Facebook was not well known inside the city hall, the most challenging part was to obtain permission from the mayor to include the project in the municipal government's budget. As the leader of the division in charge of implementing the project, the director of the Civic Activities Division bore the full brunt of securing budget at launching stage of the project. Although the staff of the Civic Activities Division prepared budget request documents, it was the director who would sit in the budget hearings to make sure they fully comprehend the project was key to a successful launch. (A)
(2) Budget and personnel constraints (A)	The use of Facebook itself did not require much of a budget initially. However, to initiate the project, first, it was necessary to compose a "project plan" and submit it to the mayor, and then it was necessary to design the banner for the Facebook page and advertise the project to local citizens. None of these tasks were covered by the initial budget, so I had to take care of them myself on a part-time basis. (A)
(3) IT skills of the staff (A, C)	I opened my Facebook account at the time of the project launch. Although reluctantly, the director of the Civic Activities Division also opened his Facebook account and showed understanding toward the project. (A) The new regular employee in charge of the project had no ICT background and had to spend most of the time getting himself familiar with the new job. There was little time for interviews so the frequency of having interviews dropped to one or two times per month. (C)
(4) Conflict with other job duties (C)	When the specially hired full-time community manager was not available, postings got delayed. In addition, when the regular employee in charge of the project was busy with subsidy disbursement because posting decisions must be made collectively, postings were delayed. In such occasions, two interview reports would be split into three posts. (C)
(5) Delays resulted from personnel changes (B, C)	It was my first year in the municipal government when I was assigned to the project. I barely knew the entrance to the city hall, not to say managing the Facebook page. I think it took me until June to be able to handle the job. (B) In the first months after the personnel change (April to June), "it was so tough because I had to take care of everything on my own." The system is designed so that a young staff member can be in charge of the daily operation of the page under the supervision of a managerial staff member. Hence, the project pauses whenever there is a personnel change for this position. (C)

(*continued*)

Table 5.2 (continued)

	Selected comments by the interviewees
(6) Strict decision-making process (A, B, C)	Permission from either the director or the deputy director is required to post comments or share information. Therefore, although it does not take as long as 24 hours, I think it can take as long as a half-day to respond to a comment. (B) After the specially hired full-time community manager finishes his interview report, the report is passed to the regular employee who accompanied them to the interview to check the contents. After that, the report will be submitted to the director of the Civic Activities Division for final approval. Because of such a decision-making system, there is a time lag between an interview and a post. For example, the report on an interview conducted with a social group on January 26 was posted on February 12/13 (amounting to around a two-week time lag). However, when the amount of reports is low in stock, the decision-making time can be shortened. (C)
(7) Response to irrelevant posts (B)	In order to compose a guideline on how to respond to posts for commercial purposes, I consulted with other self-government associations that also maintained Facebook communities. However, none had such guidelines and irrelevant posts were addressed case-by-case. (B)
(8) Privacy concerns (B, C)	It is necessary to contact the person who posted before sharing any private posts. (B) We are careful not to take photos that reveal personal information of the interviewee when conducting interviews. In case such photos were taken by accident, we will edit the photo before posting it online. (C)
(9) Assessment—has project has achieved initial goals? (A)	It seems that the Tsukuba Civic Activities Cyber-Square as a project has progressed smoothly. Although it was difficult at times in the beginning, we managed to secure the budget. However, we have yet to confirm whether the project has successfully revitalized civic activities and whether the number of civic activity groups has increased. As over three years have passed since the inception of the project, it is perhaps time to evaluate the project against such measures. Perhaps we have already reached the limit of promoting civic activities via Facebook. (A)

1 The names of the interviewees have already been replaced with job titles

ii. **Budget and Personnel Constraints**

Employee A also raised the issue of budget and personnel constraints as challenges that were faced at the early stage of the project. Although there was no need for a large amount of money to create a system using existing platforms such as Facebook, because there was no budget to hire full-time staff to manage the page at the beginning, the Planning Division staff had to design the Facebook page and advertise the project as a volunteers without being credited (A).

iii. **IT Skills of the Staff**

The interviewees also deemed acquiring the IT skills required to use social media like Facebook were regarded as a considerable challenge they had to face (A, C). First, those who were not good at ICT or who had *nigate-ishiki* at using computers had to "make their best efforts to familiarize themselves with the new duty," and consequently were only able to conduct one or two interviews per month for the Tsukuba Civic Activities Cyber-Square (C). In fact, most staff members of the Civic Activities Division opened their Facebook accounts only after the project was launched, including the director (A).

iv. **Conflict with Other Job Duties**

Employee C pointed out that management of the Tsukuba Civic Activities Cyber-Square could, at times, conflict with other job duties that were their main occupation. As previously described, employee C was hired as a contract employee to manage the Facebook page on a full-time basis. However, regulations require a regular full-time employee to be present at all interviews and that all decisions on web posts must be made together with a regular full-time employee. Such rules are implemented to reduce mistakes or errors in posts and maintain accountability.

As a result, when the regular full-time employees were preoccupied with other tasks, the frequency of conducting interviews could drop and posting of the information would be delayed. Specifically, the Civic Activities Division is composed of three sections or desks: the Civic Activities Desk, the Civic Collaboration Desk, and the Self-Government Promotion Desk. The Civic Collaboration Desk is chiefly in charge of the implementation of the Tsukuba Civic Activities Cyber-Square project. However, the same desk is also in charge of disbursement of subsidies to social groups (in accordance with Tsukuba City Administrative Organization Regulations Article 6). Consequently, when the regular employee was busy with subsidy disbursement, two interview reports were split into three posts when they had a deficit in procuring content (C) to save time.

v. **Delays Resulted from Personnel Changes**

The changes that result from personnel transfers could also make the operation difficult (B, C). Employees B and C took over the job from their predecessors in 2013, but there were no procedures or formal job training aside from some oral instruction and some written documents (B, C). However, at the time, employee B was still at the start of his first year as an employee in the municipal government and was "not even quite sure of which door was the proper entrance to enter the city hall." As a result, although the job had begun in April, it took employee B until June to be able to manage the tasks involved. As a matter of comparison, most successors also "had hard times" immediately after having taken over the new position.

vi. **Strict Decision-Making Process**

Whether or not to post a comment and to share some information is a decision collectively made by the director or deputy director of the Civic Activities Division and the staff in charge of interviews. A long period of time was sometimes required between an interview and that actual interview report posting due to the long decision-making process (A, B, C). In particular, it could take as long as two weeks for an interview report to be posted on Facebook when the regular employee was preoccupied with other jobs (C). At less busy times, it can still take close to a full working day (B) or—two to three days (C) to decide whether a post should be made. Nevertheless, all three interviewees agreed that, although such a decision-making system can take a long time, it is a necessary procedure in order to keep good job records and ensure accuracy.

vii. **Response to Irrelevant Posts**

As Facebook allows any Facebook account holder to easily post anything, posts can sometimes be placed on the Tsukuba Civic Activities Cyber-Square for commercial purposes. In order to compose a guideline on how to respond to such irrelevant posts, employee B consulted a number of self-government associations that also maintain Facebook communities. However, to the extent of the employee's knowledge, employee B found that none of the self-government associations had formal guidelines and irrelevant posts were addressed on a case-by-case basis. Consequently, the Tsukuba municipal government adopted a similar policy, which could lead to further delays (B).

viii. **Privacy Concerns**

Sometimes photos that could potentially reveal personal information were attached to posts by citizens and citizens' groups. Such cases raised specific concerns in relation to privacy and personal information (B, C). Employee B always contacted the person who uploaded the post directly to confirm sharing permissions before sharing a post through the government site. Such confirmation might cause a delay for "a certain amount of" time. Some think such confirmation to be unnecessary, but employee B said that because of the public nature of a Facebook page and the number of readers, it was necessary to exercise caution with information shared. In addition, photos of children were to be edited prior to posting in order not to reveal any personal information (C).

ix. **Assessment of Whether the Project Has Achieved the Initial Goals**

Employee A, who was involved in the launching of the project, was in charge of reviewing whether or not the project had achieved its initial goal to promote connections and communications among civic activity groups, citizens, and the government. Unfortunately, no such reviews have been carried out to date. Overall, the Tsukuba Civic Activities Cyber-Square project has progressed smoothly, with the project being added to the annual municipal government budget and beginning to hire full-time community managers (A). However, given that the project has been launched and has been progressing over three years, it is ripe to assess whether it has actually achieved the goal of revitalizing civic activities in Tsukuba via Facebook (A).

In this section, we gained important information through conducting semi-structured interviews to identify the difficulties in Facebook use faced by local governments. As described here, we were able to identify nine such difficulties, which support our experience-based assumptions. We believe the identified difficulties can be grouped into the following three larger categories.

Category 1: Quantity and Quality of Resources Involved in Project Operation
The first category of problems is associated with the quantity and quality of resources necessary for utilizing social networking services, such as budget and staff. Because the Tsukuba Civic Activities Cyber-Square uses the already-existing platform of Facebook, it generates lower launching and operating costs when compared to the custom and expensive *Chiiki SNS*, which requires the construction of original platforms and servers. Nevertheless, there are still costs related to allocating human resources to plan and lead the launching of the project and to interact with local citizens on behalf of the government, such as to conduct interviews. In addition, hiring full-time community managers will also generate costs. While the existence of full-time community managers is key to operational efficiency, they are still hired as contract employees. As long as these people are hired on a contract basis and are required to work alongside regular employees, there will inevitably be conflicts between community management and other job duties (Nakano 2014a). Hence, in order to ensure efficient operation of the Tsukuba Civic Activities Cyber-Square, it is necessary not only to hire full-time community managers but also to reconsider the impact of employment and operation systems.

The low ICT efficacy or *nigate-ishiki* of ICT among the Civic Activities Division staff is also a problem. While those who use Facebook in their private lives are familiar with the platform and are aware of both its opportunities and pitfalls, those who do not are required to start by opening personal accounts. With respect to the latter case, our interviews have confirmed that *nigate-ishiki* of ICT could lead to delays in fulfilling job duties. Therefore, quantity and quality of budget and human resources will affect the utilization of SNSs.

Category 2: Red Tape
Secondly, red tape in administrative activities has also caused trouble for community managers. Specifically, the interviewees have pointed out that whether staff members of the administrative division in charge of implementation have fully comprehended the purpose of the project, periodic personnel changes and strict decision-making procedures can both complicate the job and lead to delayed responses. In particular, the system of periodic personnel changes means that the delays resulting from personnel changes will keep recurring.

Category 3: Privacy Concerns
Thirdly, concerns about the protection of personal information and how to respond to improper posts can sometimes also cause trouble. The awareness that the public entity is responsible reinforces such concerns. For example, as a government employee, employee B gave clear expression to an obligation not to invade the privacy of citizens by saying, "precisely because the Tsukuba Civic Activities Cyber-Square is a government-run platform, we should never make any mistake." Such concerns, however, sometimes can cause delays.

Do Advertisements Actually Decrease Online Activity?

As discussed in the first section, various inconveniences in relation to maintaining a social networking service page occur because of the customs embedded in Japanese local governments. To overcome low participation and to facilitate recognition and awareness of the existence of the Tsukuba Civic Activities Cyber-Square Facebook page, the Tsukuba municipality used paid advertisements available through the Facebook community page interface.

In this latter section of this chapter, we focus on the government's creation of an online intersection between pre-existing civil society organizations and offline individuals. This section explores how actions taken by local governments influence trend fluctuations in the online engagement of Japanese civil society on an SNS page. This section also discusses how some government initiatives increase engagement but at other times, the use of paid advertisements on SNSs such as Facebook negatively affected engagement.

The following research questions were established to examine the fluctuation of engagement related to online interaction on SNSs among citizens, civil society, and government. This section explores the potential drivers (social gatherings) and barriers (financial resources such as paid advertisements) of social networking pages in nurturing civil society regarding these fluctuations in engagement.

RQ1: How do targeted online advertisements and organized offline gatherings alter communication and participation on Facebook?
RQ2: Why does the use of advertisements on Facebook not increase engagement among citizens?

To understand the dynamism of engagement on Facebook, we will employ a time series analysis and a social network analysis. Insight data including engagement, comments, impressions, and so on from the Facebook insight page will be used for both analyses.

Time Series Analysis: The Relationship Between Ads on Facebook and Engagement

We divided the timeline of the Facebook page into three phases (categorized by five time divisions) according to the content of the activities. We compared the level of civic participation in each phase. With regard to the three phases, the first phase is a period without paid advertisements or offline gatherings. In other words, we did not conduct any artificial experiments during this phase, and the Facebook page at this time was being managed spontaneously. Specifically, November 1–26, 2014 (①) and March 21–April 30, 2015 (⑤) are categorized as the time divisions in this particular phase.

The second phase involved the period after the initiation of offline gatherings on the Tsukuba Civic Activities Cyber-Square Facebook page. Specifically, December 28, 2014 through March 4, 2015 (③) is the time division for this phase.

The third phase involved the initiation of paid advertisements on Facebook. Specifically, November 27, 2015 through December 27, 2015 (②) and March 5, 2015 through March 20, 2015 (④) are the time divisions for this phase. The geographic location of the advertisements included citizens and organizations registered in Tsukuba, Ibaraki, who were targeted through the facilitation of employing an advertising agent because direct, paid advertisements were not to be approved by any of the entities that were involved with this initiative (Table 5.3).

This study uses the following two benchmarks which indicate civic participation as obtained from the Facebook Insights menu, which is available in the interface for community pages.

- The term "reach" is used to describe the total number of people who have seen any content associated with the page.
- The Facebook insight term "engagement" is used to indicate the number of people that interacted with the page per month, inclusive of any time a person clicked on or created a story.

Table 5.3 Three phases and five time divisions

	Phases	*Time divisions*
1	Operation of the Facebook page without ads or offline gatherings	November 1–26, 2014 (①) March 21–April 30, 2015 (⑤)
2	After offline gatherings were initiated	December 28, 2014–March 4, 2015 (③)
3	Period when paid ads on appeared on the Facebook page	November 27–December 27, 2015 (②) March 5–20, 2015 (④)

"Reach" describes the total number of citizens who have seen any content associated with the page, inclusive of advertisement and posts by others. "Engagement" in this section deviates from normal usage of the term in related literature and is limited to usage of the Facebook technical term that indicates the total number of citizens who had clicked on, commented on, or shared a post. The engagement value here is important as it indicates the number of citizens who actually participate on the page through some type of action online through the page, while the reach value is a relatively passive measurement, indicating the potential of "eyeballs" looking at the page.

Social Network Analysis: The Relationship Between Ads and the Network

The aim of this section is to identify the factors that define the level of engagement of citizens with the Tsukuba Civic Activities Cyber-Square. In an attempt to understand these factors, we will analyze and describe the networks created by the municipal government and citizens and compare the network structures based on the phases identified above. Indicators of the closeness centrality can be used to measure actors' levels of interaction with each other in the network. Closeness centrality and betweenness centrality indicate patterns in how actors make ties. These centrality measures can be used to see the ties among those engaged in the network (Borgatti 2005).

- Closeness centrality can be used to measure how much an actor directly communicates with other actors.
- Betweenness centrality can explore the degree to which an actor plays a bridging role among other actors. The bridging role here is critical as this is the focal point where important information flows

from the leading non-profit organization to others. This will lead to subsequent information flows to other followers or members in subgroups.

The visualization of the network allows us to see the mutual communication and engagement patterns that are involved on the Tsukuba Civic Activities Cyber-Square Facebook page. In conducting our analysis, we made use of Node XL, which is a software package that allows us to analyze mutual networks and display graphs of SNSs such as Facebook or Twitter. In addition, the clusters can be used to identify the subgroups of a particular Facebook page. These results will provide evidence regarding the patterns of engagement on the Facebook page. The network is visualized based on the comments created by citizens and groups on the Facebook page. Vertices indicate citizens on the Tsukuba Civic Activities Cyber-Square Facebook page. Edges will be created by: (1) users who commented on the same post; (2) commenters and post authors; (3) two consecutive commenters; (4) commenters and comment authors; and (5) post/comment authors and users tagged.

Relationship Between Ads on Facebook and Engagement

As mentioned earlier, we divided the timeline of the Facebook page into three phases (categorized as five time divisions) based on the content of the activities. We compared the level of civic participation using values for "engagement" and "reach." The results for each of the three phases are presented below.

Operation of the Facebook page without ads or offline gatherings (①, ⑤): Stable engagement and stable reach

Figure 5.1 indicates a shift in both engagement and reach from November 2014 through April 2015. One can observe how engagement and reach are stable from November 1 to 26, 2014 (①) and from March 21, 2015 through April 30, 2015 (⑤).

1 Left axis indicates daily total reach, and right axis indicates daily page engaged users. After offline gatherings were initiated (③): Relatively high engagement and stable reach

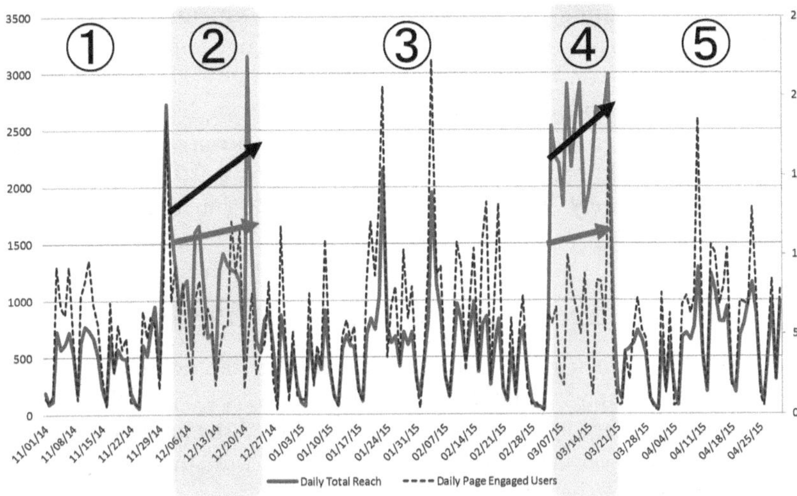

Fig. 5.1 Total reach and engagement values for citizens visiting the Tsukuba Civic Activities Cyber-Square Facebook page

As mentioned above, social gatherings were organized to promote the usage of the Tsukuba Civic Activities Cyber-Square on Facebook and to encourage networking among citizens. Before the offline gatherings, the local government had posted news related to events on the Tsukuba Civic Activities Cyber-Square Facebook page and attempted to enhance communication among citizens as well.

After offline gatherings were initiated, one can observe how engagement increases, especially between January 2015 and February 2015. On the other hand, the reach value is approximately 1,000–2,000 per day and is relatively stable.

Period when paid ads appeared on the Facebook page (②, ④): Extremely high reach and stable engagement

The black arrow indicates a data shift of the reach value of the Tsukuba Civic Activities Cyber-Square page during and after paid advertisements, which appeared from November 27 to December 27, 2015 (②) and from March 5 to 20, 2015 (④). During these periods, the reach values increased rapidly to approximately 3,000 per day and were relatively stable at moderate levels during the period of paid advertisements. On the other hand, the gray arrow represents a data shift in engagement during the same period. One can observe that the engagement level is not strongly related to the period of paid advertisements. From November 27 to December 27, 2015 (②), the engagement level is around 30 to 150 per day and does not fluctuate much. From March 5 to 20, 2015 (④), engagement is approximately 20–180 per day and does not fluctuate much here either.

The Relationship Between Ads and the Network

The next aim of this study is to identify the factors that define the level of citizen engagement with the Tsukuba Civic Activities Cyber-Square. To understand these factors, we analyzed and described the networks created by the municipal government and citizens and compared the network structures based on the phases discussed above.

Operation of the Facebook page without ads or offline gatherings (①, ⑤): A relatively strong network among citizens

We first describe the network analysis results of the Tsukuba Civic Activities Cyber-Square Facebook page without paid advertisements or offline gatherings. Figure 5.2 shows the network structure from November 1 to 26, 2014 (①). Figure 5.3 shows the network structure from March 21 through April 30, 2015 (⑤). The dark circle within the network link is the Tsukuba Civic Activities Cyber-Square.

A mutual network or ties based on conversations can be observed between the municipal government and citizens. One can see that the municipal government led the conversations and communicated with citizens. At the same time, different groups were spontaneously generated among citizens, and they interacted with each other and had conversations within those groups. As a result, a relatively strong network or strong ties can be observed not only between the municipal government and citizens but also among citizens in the same group.

After offline gatherings were initiated (③): Appearance of a leader

Figure 5.4 shows the network structure after offline gatherings were initiated, in the period from December 28, 2014 through March 4, 2015 (③). A mutual network or ties based on conversations can be observed between the municipal government and citizens. At the same time, different

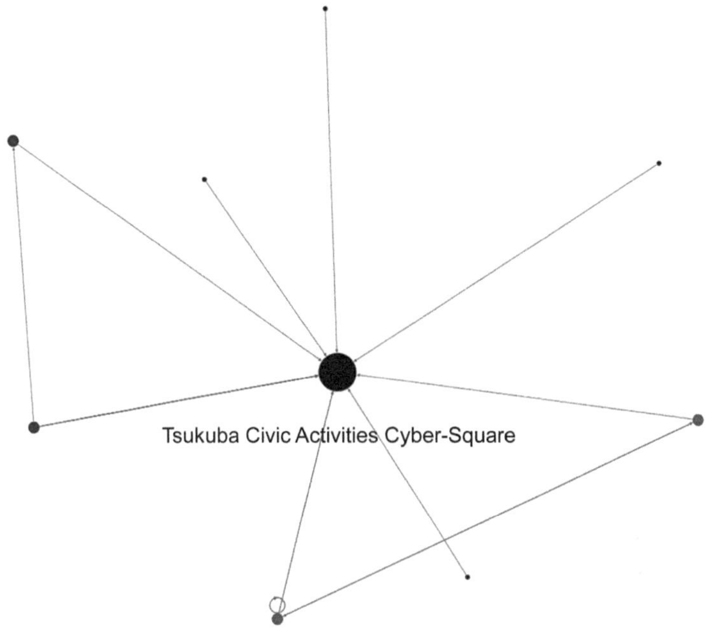

Fig. 5.2 Node XL analysis result map (①)

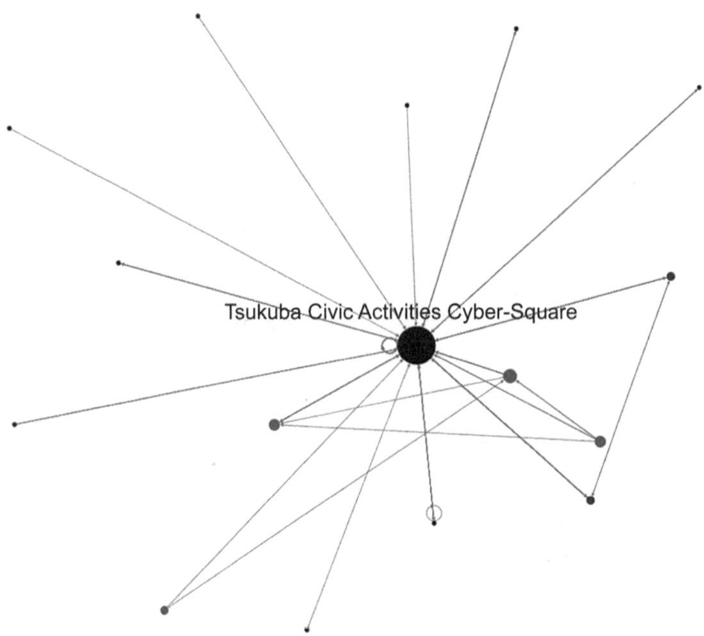

Fig. 5.3 Node XL analysis result map (⑤)

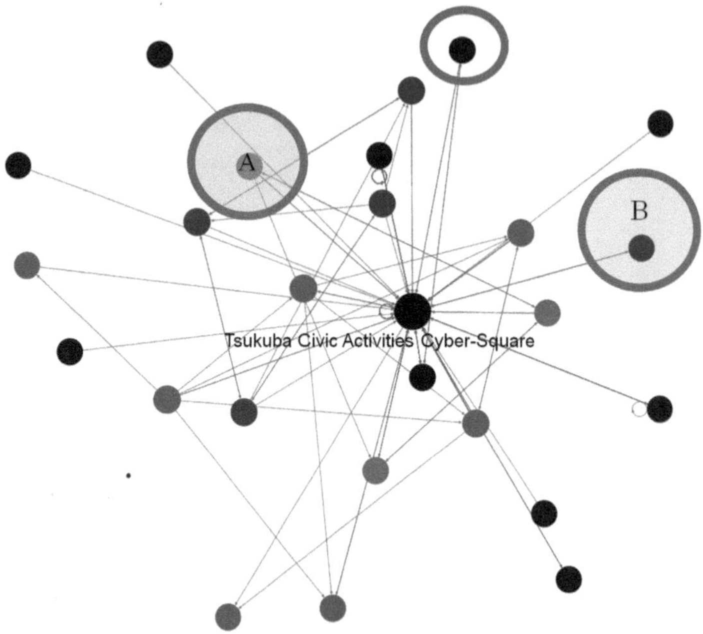

Fig. 5.4 Node XL analysis result map (③)

groups were generated spontaneously among citizens, and they interacted with each other and had conversations within those groups.

The encompassing gray circles within the network indicate participation in an offline gathering. One can see that the citizens who participated in the offline gatherings led the conversations within the group. For example, citizen A of Fig. 5.4 in the gray circle is a leader of an educational group located in Tsukuba who had lively conversations with the municipal government. Citizen B of Fig. 5.4, who is a member of a welfare group, had conversations not only with the participants in offline gatherings but also with citizens who did not participate in the gatherings. We found that a leader of a neighborhood association expressed gratitude for the daily support of municipal hall workers, and the workers also expressed appreciation for the kind words and encouraged group members to be more active. Specifically, the welfare group posted the comment, “Thank you so much for your kind support of our activities,” and the municipal hall worker responded by saying, “I should say thanks to you, likewise. Looking forward to your success in the next academic year.” In another case, a member of an educational group made a comment on the newsfeed stating, "It was a great opportunity for us to join the event and make new network connections here in Tsukuba," and the member encouraged the municipal government to create more opportunities for citizens to communicate with each other.

As a result, a relatively strong network or strong ties can be observed not only between the municipal government and citizens, but also among citizens in Fig. 5.4. Some of the citizens who participated in offline gatherings led the conversations in the groups and the citizen networks within the same groups became stronger.

Period when paid ads appeared on the Facebook page (②, ④): Disappearance of the network

Figure 5.5 shows the network structure during the period of paid advertisements, November 27 through December 27, 2015 (②). We can

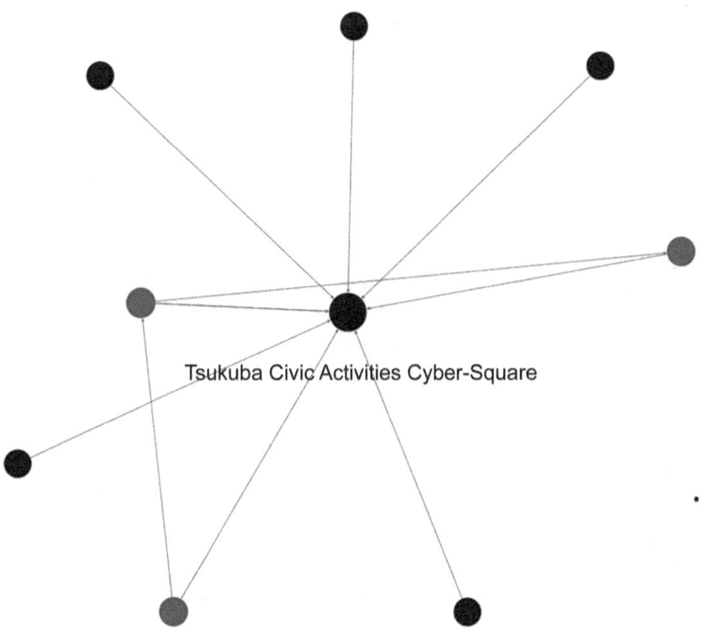

Fig. 5.5 Node XL analysis result map (②)

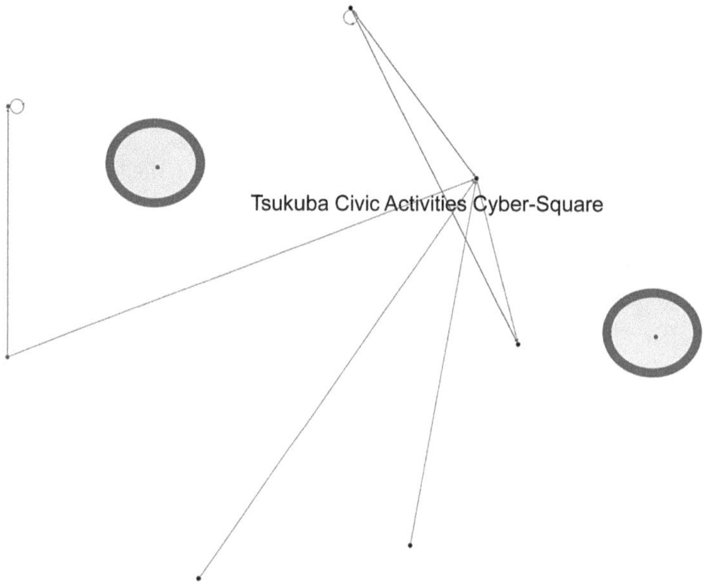

Fig. 5.6 Node XL analysis result map (④)

easily see that there is less of a network among citizens than in the phase after offline gatherings were initiated because they stopped communicating with each other even though some groups still existed in the network. To understand the reasons why the network disappeared, we need to take a closer look at Fig. 5.6, which illustrates the network structure during the paid advertisement period from March 5 to 20, 2015 (④), just after an offline gathering was held by the local government.

The circles within the network in Fig. 5.6 indicate participants in the offline gathering. We can observe here how "newcomers" to the community started conversing, while other citizens stopped their conversations and rarely participated. For example, the leader of the educational group who participated in the offline gathering stopped communicating not only with the municipal government but also with citizens in the same group. Another citizen carried on conversations only with the municipal government. As we mentioned earlier in this chapter, at the beginning of the fiscal year in Japan, local and national governments routinely shuffle personnel annually or bi-annually. The reason for these staff changes is to avoid collusion that occurs when workers become too comfortable with their assignments. Another reason for this inefficiency is to create well-rounded employees. Although this may seem a reasonable rationale for annual personnel reassignments, this tradition of shuffling people around each April creates lapses in efficiency all throughout Japan because new workers are assigned to new tasks every April. Thus, one reason for the disappearance of the network may be that municipal workers who were in charge of the Tsukuba Civic Activities Cyber-Square Facebook page were reassigned during this period and the new staff members did not pay much attention to the page.

Summary of Results

Our time series analysis and the analysis of network shapes among the citizen links were generated as follows (Table 5.4). We divided the timeline of the Facebook page into three phases (separated into five time divisions) and confirmed citizen engagement using the "reach" and "engagement" values of the Facebook Insights interface. The next aim of this section was to identify the factors that define the level of citizen engagement with the Tsukuba Civic Activities Cyber-Square. To understand these factors, we described the networks between the municipal government and citizens and compared the structure of the networks based on the three phases.

The first phase was the Facebook page during the period without ads or offline gatherings from November 1 to 26, 2014 and March 21, 2015 to April 30, 2015. One can observe how both engagement and reach are stable during this period. At the same time, a relatively strong network or relatively secure ties can be observed not only between the municipal government and citizens but also among citizens in the same group, because the municipal government led the conversations and maintained the public sphere on the Tsukuba Civic Activities Cyber-Square Facebook page.

The second phase was after the offline gatherings were initiated from December 28, 2014 to March 4, 2015. During this period, one can see that engagement generally increased while the reach values were relatively stable. At the same time, a relatively strong network or strong ties can be observed not only between the municipal government and citizens but also among citizens in the same group, because some citizens who participated in the offline gatherings led conversations in the groups and the networks among citizens within the same groups became stronger.

Table 5.4 Comparison

	Phases	*Time divisions*	*Reach*	*Engagement*	*Network*
1	Operation of the Facebook page without ads or offline gatherings	November 1–26, 2014 (①) and March 21–April 30, 2015 (⑤)	○	○	• A mutual network and conversations among citizens can be observed • Municipal governments led the conversations. Some different groups were spontaneously generated • A relatively strong network/ties can be observed not only between the municipal government and citizens but also among citizens in the same group
2	After offline gatherings were initiated	December 28, 2014–March 4, 2015 (③)	○	◎	• A mutual network and conversations among citizens can be observed. Some different groups were spontaneously generated • Some citizens led the conversations in the groups and networks among citizens in the same groups became stronger
3	Period when paid ads on appeared on the Facebook page	November 27–December 27, 2015 (②) and March 5–20, 2015 (④)	◎	○	• "Newcomers" to the community started conversing while previous members stopped conversing • Networks among previously engaged citizens barely existed. (Citizens stopped carrying on conversations) • Some citizens carried on conversations only with the municipal government

1 ○=stable, ◎=increase

The third phase was the period after paid advertisements appeared on the page from November 27, 2015 to December 27, 2015 and from March 5 to 20, 2015. During this period, the reach values increased rapidly while engagement was relatively stable. This is because "newcomers" started conversing while the previous members stopped carrying on conversations and networks and groups barely existed.

Our findings indicate that the different stages of growth can be attributed to a blend of periodic social gatherings and paid advertisements. Social gatherings were found to be more beneficial for engagement because they can create leaders in the network among citizens. On the other hand, paid advertisements may be less effective for engagement, because the existence of paid advertisements seems to result in diminished citizen networks.

Discussion

The importance of the growth of cooperation between the public and private sectors has been emphasized greatly because political participation through cooperation can create democratically minded citizens who are more aware of public issues. One of the ways to mediate issues between the government sector and the private sector is through the use of social media such as Facebook. Social media has the potential to create useful environments for citizens so they can participate virtually and actively toward a more open government. Citizens can become the producers of public services through social media. Opportunities created by enhanced online communication allow for the creation of greater social capital and a more effective civil society.

The aim of the first section was to identify the various difficulties faced by employees of local governments when using a social networking service to share information. The authors conducted exploratory interviews with three employees of the Tsukuba municipal government who were involved in the operation of the municipal government's official Facebook page—the Tsukuba Civic Activities Cyber-Square. The interviews revealed that nine factors could inhibit its operation: (1) appreciation and support from the administrative division in charge; (2) budget and personnel constraints; (3) staff IT skills; (4) conflicts with other job duties; (5) personnel changes; (6) strict decision making process; (7) response to irrelevant posts; (8) privacy concerns; and (9) whether project goals have been achieved. These factors can be grouped into three categories: the quantity and quality of resources involved in project operation, red tape, and privacy concerns.

In the second section, to understand the dynamism of engagement on the Facebook page, we employed a time series analysis and a network analysis. Facebook Insight data including engagement, comments, impressions, and so on from the Facebook insight page was used for both analyses. We divided the timeline of the Facebook page into three phases (which were separated into five time divisions), and confirmed citizen engagement using "reach" and "engagement" values. The next aim of this study was to identify the factors to define the level of citizen engagement with the Tsukuba Civic Activities Cyber-Square. To understand the factors, we described the networks between the municipal government and citizens and compared the network structures in each phase.

Our findings indicated that the different stages of growth can be attributed to a blend of periodic social gatherings and paid advertisements. Social gatherings were found to be more beneficial for engagement because they can create leaders in the network among citizens. On the other hand, paid advertisements may be less effective for engagement, because this made the network stand out. In other words, daily communication and offline gatherings for citizens are more effective than online advertisements to create networks and expand the public sphere.

Finally, we would like to discuss possible directions for future research. This study is a pilot small-number case study on the management staff's involvement in SNS utilization. Therefore, in order to generalize findings from this study, it would be necessary to expand the sample size. In this study, we selected and interviewed employees who were actually involved in the daily operation of the Facebook page "Tsukuba Civic Activities Cyber-Square," such as interviewing local civic groups and posting interview reports. However, we did not interview any people who held the managerial posts of desk chief or higher. How these people view this project, and, closely related, how do their actions and policy preferences affect the project are subjects of future research. In addition, it is necessary to expand the number of cases studied. This study focused on the single case of the Tsukuba Civic Activities Cyber-Square which was operated by the Tsukuba municipal government. However, whether or not the same factors/difficulties will be identified in cases of their local governments is a question that requires further investigation. Hence, we are considering it as another subject of future research.

References

Borgatti, S. P. (2005). Centrality and network flow. *Social Networks, 27*(1), 55–71.

Borry, E. (2013). *Red tape: A new measurement.* Paper presented at the 11th Public Management Research Conference, Madison, Wisconsin. Retrieved from http://www.union.wisc.edu/pmra2013/Paper Submissions/Renamed/RedTape ANew Measurement.pdf

Bozeman, B. (1993). A theory of government 'red tape'. *Journal of Public Administration Research and Theory, 3*(July), 273–303.

Bozeman, B., & Feeney, M. (2011). *Rules and red tape: A prism for public administration theory and research.* Armonk: M.E. Sharpe.

Cullen, R. (2008). Citizens' concerns about the privacy of personal information held by government: A comparative study, Japan and New Zealand. In *Proceedings of the 41st Hawaii International Conference on System Sciences* (pp. 224–224). doi:10.1109/HICSS.2008.91

Dahlberg, L. (2001). The Internet and democratic discourse: Exploring the prospects of online deliberative forums extending the public sphere. *Information, Communication and Society, 4*(4), 615–633.

Feeney, M. (2011, June 2–4). Organizational red tape: The conceptualization of a common measure. *Annual Public Management Research Conference Syracuse, N.Y., 4,* 1–31. Retrieved from https://www1.maxwell.syr.edu/uploadedFiles/conferences/pmrc/Files/Feeney_OrganizationalRedTapeThe Conceptualization of a CommonMeasure.pdf

Feeney, M. K. (2012). Organizational red tape: A measurement experiment. *Journal of Public Administration Research and Theory, 22*(3), 427–444. doi:10.1093/jopart/mus002.

Kabashima, I. (1988). *Seiji sanka [Political participation].* Tokyo: Tokyo University Press.

Kaufman, H. (1977). *Red tape: Its origins, uses and abuses.* Washington, DC: Brookings.

Kavanaugh, A. L., Fox, E. A., Sheetz, S. D., Yang, S., Li, L. T., Shoemaker, D. J., Natsev, A., & Xie, L. (2012). Social media use by government: From the routine to the critical. *Government Information Quarterly, 29,* 480–491.

Keane, J. (2000). Structural transformation of the public sphere. In K. L. Hacker & J. van Dijk (Eds.), *Digital democracy: Issues of theory and practice* (pp. 70–89). London: Sage Publications.

Kodagiri, Y. (2014). *Gyōsei-shiminkan kyōdō no kouyou : Jisshouteki sekkin [Cooperation between administration and citizens: An empirical approach].* Tokyo: Hōritsu Bunkasha.

Kogawa, I. (2012). Fujisawashi shimin denshi kaigishitsu 15 nen no rekishi: Sono doutai kara nani wo manabuka [The 15 years of the Fujisawa citizen electronic meetings: What can we learn]. *Journal of the Musashi Sociological Society, 14*(1), 35–54.

Lee, G., & Kwak, Y. H. (2012). An open government maturity model for social media-based public engagement. *Government Information Quarterly, 29*, 492–503.

Linders, D. (2012). From e-government to we-government: Defining a typology for citizen coproduction in the age of social media. *Government Information Quarterly, 29*, 446–454.

Merton, R. K. (1949). *Social theory and social structure: Toward the codification of theory and research*. Washington, DC: Free Press.

Nakano, K. (2014a). Kansei chiiki SNS ga haishi ni itaru keii ni kansuru kousatsu: Jichitai shokuin eno intabyū chōsa yori [Why the Chiiki SNS were abolished: Interviews with local government employees]. *Social and Economic Systems Studies, 35*, 35–42.

Nakano, K. (2014b). Chiiki SNS heno chihou jichitai shokuin no kanyo jittai ni kansuru kousatsu [The involvement of local government employees with Chiiki SNS]. *Socio-Informatics, 2*(3), 1–14.

Nishio, M. (2004). Hashigaki [Foreword]. In T. Nishio (Ed.), *Jūmin, comyunityī tono kyōdō. [Cooperative work of residents and community]* (p. iii). Tokyo: Gyousei.

Noguchi, M., & Ito, N. (2013). Jichitai ni okeru Facebook kouhou ni kansuru medhia kouka sokutei: Sagaken takeoshi no sōsharu kyapitaru to shibikku pawā heno eikyō [Media effect measurement of local government public relations through Facebook: Effects of civic power and social capital Takeo, Saga]. *Journal of the Japan Information-Culture Society, 20*(2), 35–42.

Okura, S., & Kaigo, M. (2016). Who leads advocacy through social media in Japan? Evidence from the "Tsukuba Civic Activities Cyber-Square" Facebook page. *Information, 7*(4), 66. doi:10.3390/info7040066.

Poster, M. (1995). *The second media age*. Cambridge: Polity Press.

Riccucci, N. (2012). Reconceptualizing red tape research. *Public Administration Review*, 463–465. Retrieved from http://onlinelibrary.wiley.com/doi/10.1111/j.1540-6210.2011.02552.x/abstract

Sakamoto, H. (2010). *Sōsharu kyapitaru to katsudou suru shimin: Shin jidai nihon no shimin seiji [Social capital and civic activities: Citizen politics and a new era of Japan]*. Tokyo: Yuhikaku.

Sassi, S. (2000). The controversies of the Internet and the revitalization of local political life. In K. L. Hacker & J. van Dijk (Eds.), *Digital democracy: Issues of theory and practice* (pp. 90–104). London: Sage Publications.

Shoji, M. (2008). Chiiki SNS saito no jittai haaku, chiiki kasseika no kanousei [Understanding the situation of Chiiki SNS, and the possibilities of regional reactivation]. *Jōhō tsūshin seisaku kenkyū puroguramu kenkyū seika ronbun*, 1–38. Retrieved from http://www.officepolaris.co.jp/icp/2007paper/2007014.pdf

Shoji, M. (2012). Chiiki ni okeru shakai nettowāku to jōhō tsūshin gijutsu [Information and communication technology and social networks in regions]. In S. Sugimoto (Ed.), *Jōhōka jidai no rōkaru komyunityī: ICT wo katsuyou shita chiiki netowāku no kouchiku [Local communities in the age of information: Construction of local networks through ICT]* (pp. 61–80). Osaka: National Museum of Ethnology.

Tanaka, T., & Yamaguchi, S. (2016). *Netto enjō no kenkyū: Darega aori, dou taisho surunoka [Net enjyo research: Who does it and who is it aimed at]*. Tokyo: Keisou Shobou.

Wilhelm, A. G. (2000). *Democracy in the digital age: Challenges to political life in cyberspace*. New York: Routledge.

Muneo Kaigo is an associate professor in the Faculty of Humanities and Social Sciences at the University of Tsukuba in Japan. He teaches courses in media communication in the Graduate School of Humanities and Social Sciences, media management in the Graduate School of Business Sciences and communication science in the College of Comparative Culture at the University of Tsukuba. He is currently leading a joint research project on social media uses among local municipalities in Japan with cooperation from the municipal government of Tsukuba and Intel Corporation Japan.

Sae Okura is an assistant professor in the Faculty of Humanities, Law and Economics at Mie University in Japan. She conducts research in politics and social policy studies. Her recent research interests focus on political participation by minority groups such as the disability community, sexual minorities, and single parent families. She analyzes quantitative survey data collected in Japan.

CHAPTER 6

Who Leads Advocacy Through Social Media in Japan?

Sae Okura and Muneo Kaigo

What Is Advocacy and Why Is It Important?

The importance of advocacy activities by civil society organizations (CSOs), including public interest groups, citizens' groups, non-profits, neighborhood associations and social movement organizations, in policy and decision-making procedures has been greatly emphasized in the political science and social policy literature (Child and Grønbjerg 2007; Salamon and Geller 2008). Salamon and Geller (2008) indicated, for example, that "active participation in the policy process is a fundamental function of the non-profit sector in a democratic society." Child and Grønbjerg (2007) have noted that advocacy activities by CSOs might be the "primary vehicles by which people … pressure government to respond to disadvantaged groups … and attend to unresolved problems." Pekkanen and Smith (2014) also pointed out that attention to advocacy by CSOs was growing because several valuable contributions have come from surveys in recent years. At the same time, Pekkanen and Smith (2014) pointed

S. Okura
Faculty of Humanities, Law and Economics, Mie University, Tsu, Mie, Japan

M. Kaigo (✉)
Faculty of Humanities and Social Sciences, University of Tsukuba in Japan, Tsukuba, Ibaraki, Japan

M. Kaigo (ed.), *Social Media and Civil Society in Japan*,
DOI 10.1007/978-981-10-5095-4_6

out that advocacy behavior generally resisted scholarly analysis for some critical reasons. That is, it is difficult (1) to measure advocacy because it covers a broad range of actions and (2) to determine its causality.

At the same time, local and central governments are motivated to focus on the CSOs. After World War II, the demands from citizens for social services regarding problems related to childcare or nursing care, environmental problems, and employment measures, have increased and diversified. On the other hand, since the 1970s, the government has been withdrawing from spending on public policy or reducing its financial role as the result of financial stringency and the influence of the policy of neoliberalism. In other words, while it becomes difficult for the government itself to correspond to social needs, it becomes necessary on a trial basis for the public sector to perform the role it has been taking on until now in collaboration with the private sector.

As we have already discussed, one of the ways to integrate the government sector with the private sector is through social media, such as Facebook. Social media enables an open government (Lee and Kwak 2012) and expands possibilities for residents to become more involved in the provision of public services and have more influence on policy decisions (Linders 2012). In addition, by connecting government and citizens via social media, there is an expansion of public space (Wilhelm 2000; Sassi 2000; Keane 2000; Dahlberg 2001), and social media can become a path for information and knowledge related to public issues such as citizens' daily living (Kavanaugh et al. 2012).

Despite the importance of their representative activities, both the relevance and impact of the leading actors who structure the diverse networks and discourses through social media need further recognition in both fields. The purpose of this chapter is to analyze CSOs at the local government level involved in advocacy activities through the use of social media such as Facebook and Twitter. Despite some different definitions of "civil society," we have used the conception of "civil society" defined by Schwartz (2003) as the "sphere intermediate between family and state in which social actors pursue neither profit within the market nor power within the state." This chapter again employs findings from the Japanese Facebook community page, the "Tsukuba Civic Activities Cyber-Square," run by the local government in Tsukuba City in Ibaraki Prefecture, Japan.

Advocacy Activities by Civil Society Organizations in Japan

The notion of "advocacy" has a wide range of contemporary meanings, and it is often used interchangeably with similar words such as "lobbying" and "political activity" by various authors in the bridging fields involved in studying advocacy. As mentioned earlier, Pekkanen and Smith (2014) pointed out that advocacy behavior generally resisted scholarly analysis for some critical reasons. That is, it is difficult to (1) measure advocacy, because it covers a broad range of actions and to (2) determine its causality.

Some researchers use the term "advocacy" to describe what others call "lobbying" (Bass et al. 2007). Leech (2010) argues that a broader definition of lobbying is necessary. On the other hand, Salamon and Geller (2008) introduce an analytical distinction between "policy advocacy" and "lobbying." According to their definition, "policy advocacy" is the more general term which "aims to influence government policy at the federal, state, or local level and can encompass a range of activities, including conducting research on public problems, writing op-ed pieces on issues of public policy, building coalitions, or participating in a group working to formulate a position on a matter of policy." Lobbying is a specific subset of policy advocacy and involves the communication of the organization's positions to policy-makers, either directly (direct lobbying) or by mobilizing the general public (grassroots lobbying). Salamon and Geller (2008) have also pointed out that the key difference between lobbying and other forms of advocacy is that lobbying involves taking and promoting a position on specific legislation. Bass et al. (2014) have pointed out that lobbying and advocacy are often regarded as synonymous, but advocacy is a broader concept that involves lobbying.

As the purpose of this chapter is not the creation of a new definition for "advocacy," it may be more useful to define advocacy by the activities that a non-profit engages in to influence public policy, either directly or indirectly (Salamon and Geller 2008). Furthermore, "advocacy" does not influence policy change by itself, but the "activities" that a non-profit engage in are an attempt to influence public policy. Therefore, we focus not on influence or policy change, but the communication activities such as requests by the civil society organizations directed toward the municipal government on Facebook.

Who Advocates? Japan's "CSOs" Through a Comparative Perspective

The appearance of Japan's civil society differs slightly depending on whether you compare it internationally or confirm details of the organizations within the country. This chapter will review some of the noteworthy or special features of Japan's civil society that are relevant.

First, there are some special or noteworthy characteristics of Japan's civil society that have been observed from the viewpoint of international comparison. When the special characteristics of Japan's civil society are described, resources have been found to be a significant problem. Robert Pekkanen (2006) indicates how Susan Pharr has coined the phrase "four smalls" when describing the special characteristics of Japan's civil society at the Association for Asian Studies Annual Meeting in 2005. The expression "four smalls" refers to small membership, small numbers of professional staff, small budgets, and small geographic scopes. Leng (2015) also makes references to Pharr's comment on this "four smalls" issue in his discussion of Japan's civil society. Pekkanen (2006) investigated why Japan's civil society is composed of many small groups but few large professionally managed national organizations from a political institutional perspective. He demonstrates that political institutions such as regulatory frameworks, financial flows, and political opportunity structures are the main factors underlining civil society in Japan. In his conclusion, Japan is lacking in advocacy activities that bring about social change, and he has labelled this "Members without Advocacy" (Pekkanen 2006). Pekkanen has made a comparison of the number of employed personnel of civil society organizations in Japan with other nations. According to his analysis, Japan was found to have the second smallest number of organizations employing personnel of all the OECD countries, with only Germany having a smaller number (Pekkanen 2006). Furthermore, previous study results also confirmed that Japan's CSOs have fewer employed staff than CSOs in the USA, Korea and Russia (Tsujinaka et al. 2010).

On the other hand, the survey results focused on Japan reveal that producer groups, such as economic groups, labor unions and agricultural organizations, and political groups, are more dominant in advocacy while welfare groups, educational groups, civic organizations, and professional groups are more passive in voicing their interests (Tsujinaka et al. 2010; Tsujinaka and Mori 2010). Yamamoto (2010) analyzed the data of the Second Cross National Survey on Civil Society Organizations and Interest

Groups in Japan, (J-JIGS2), and indicated that economic groups and agricultural organizations have more direct access to political elites such as the Liberal Democratic Party (LDP), bureaucrats, and local government workers, while political groups, labor unions, and civic organizations have less access to political elites and have attempted to access them indirectly via media, public opinion, courts, etc. (outside lobbying). For instance, Okura (2013) analyzed the same data, and discovered that, in regards to welfare groups, the establishment of service-type organizations has been increasing since 1996, especially at the local level, because of institutional welfare reform in recent years, while since the 1980s the establishment of advocacy-type organizations has not been increasing at both the national and local levels. (Needless to say, there are some examples in which civil society organizations have some influence in policy-making.)

Path dependency theory could explain this feature of political structure in Japan. After World War II, countries categorized as capitalist developmental states such as Japan had been protecting producer sector associations such as economic groups, labor unions, and agricultural organizations to achieve economic growth (the so-called convoy system). In this process, stronger information and human networks have been built among those associations, the bureaucracy, and the LDP in Japan, which had held continuous political power since its founding in 1955 until their loss to the Democratic Party of Japan prior to the Great East Japan Earthquake. In contrast, the non-profit and civic sectors excluded from this convoy have been discouraged from appearing in the focal areas of the Japanese political network. This older political structure (*kyū kouzou*) can be observed in a survey conducted in 2006–2007 and is a special and noteworthy characteristic of Japan's civil society (Tsujinaka and Mori 2010).

In addition, the survey results also indicate routes of information flow available through the traditional community—neighborhood associations (*Jichikai* or *Chonaikai*). The survey results clarified that neighborhood associations, rather than political parties or politicians, try to exert influence over policies by directly contacting the administration (Tsujinaka and Ito 2010). Furthermore, previous study results also confirmed that groups with capital resources such as employed personnel, budgets, and several individual members actively carry out activities in comparison to groups without resources (Yamamoto 2010).

Through these previous studies, we can summarize that Japan's civil society advocacy ability is assessed as vulnerable, based on the number of employed personnel when compared to the numbers in other nations.

Based on the previously mentioned typology of area of policies, groups such as welfare groups, educational groups, civic organizations and professional groups have fewer opportunities to make direct contact with political actors. On the other hand, Japan's civil society organizations have obtained local routes of information flow through economic groups, labor unions and agricultural organizations, political groups and neighborhood associations.

Advocacy and Media in Japan

Civil society organizations with limited resources in Japan have traditionally gained the help of mass media to communicate their messages. Around the end of the LDP-dominated era of Japanese politics of the 1970s, views of Japan's political system as being more pluralistic began to emerge. Among these views were the referent pluralism championed by Kabashima and Broadbent (1986) who emphasized the influence and the important role played by mass media.

Using data from the March 1980 study "Elites and the Idea of Equality" and citing a deep relationship between the class structure of influence as acknowledged by the elite and the actual class structure, Kabashima illustrated that: (a) there was a high degree of coincidence in the ranking of evaluations of each group's influence; and also that (b) all the group leaders, with the exception of the mass media leaders, saw the mass media as the most influential group in terms of Japan's overall political and social systems. Kabashima also underlined the two principles of: (a) neutrality, i.e., mass media leaders were politically neutral in all areas of political party support, political ideology, view of socio-economic equality, and traditional values; and (b) inclusiveness, i.e., in examining the relationship between the influence of interest groups and the degree of contact with the following four influence groups: (a) LDP leaders, (b) the bureaucratic elite, (c) opposition leaders, and (d) the mass media. The mass media had direct connections with a variety of group leaders that went beyond differences in size and the relative newness of group and political ideology, so Kabashima argued that the mass media was positioned outside the core of the traditional authority group comprising the LDP and the bureaucratic elite, and served to inject into the political system the preferences of groups that tended to be excluded from the main authority (Kabashima and Broadbent 1986; Kabashima 1990, 2004; Kabashima et al. 2007).

Recently, social media has been added alongside mass media as a way of transmitting certain information from the civil society organizations into the political system. To reiterate a point made previously, social media is one way to connect civil society organizations with local and central governments that are struggling with limited budgets. On the other hand, Kobayashi (2013) analyzed the result of an Internet survey conducted in 2012, and discovered that the proportion of respondents who recognize the partisan nature of newspapers, TV news, and Internet news is 28 percent, 18 percent, and 13 percent respectively. The result of the analysis indicates that Japanese citizens might not recognize the partisan nature of their information sources when they are viewing or reading news in normal everyday situations (Kobayashi 2013). Ogasawara (2014) analyzed aspects of social media usage and its influence on political interest, cognition of a number of important issues, and attitudes to political parties after the 2013 Japanese Upper House election. The results of his analysis indicate that: (1) only 18.3 percent of the respondents had accessed websites or social media for electoral information, while over half of them watched TV programs or read newspapers; and (2) social media usage of respondents is weakly related with political interest, cognition of a number of important issues, and attitudes to political parties. Furthermore, as online election campaigns have been allowed in Japan since 2013, attention to political and policy-related Internet usage, including social media and civic engagement, is greater (Yamazaki 2015). With regard to Japan, Tsuda (2012) also pointed out that the anti-nuclear demonstration held in front of the prime minister's official residence in 2012 was enhanced by social media usage. Okamoto et al. (2015) analyzed the impact of the internet campaigning liberalization on voters in 2013 during the upper house election held in Japan that year. Compared with the upper house election in 2010, respondents in 2013 tended to vote for candidates and accessed the websites of the candidates significantly more.

Social media users can be considered to be positioned outside the traditional authority groups, but they can create flows of information, and one can observe such changes in the behavior patterns of civil society organizations. Such changes, however, have been the subject of little discussion until recently; therefore, this chapter will attempt to observe and measure these new changes in information flows.

As mentioned earlier, previous studies have suggested that Japanese civil society organizations with limited financial and political resources have difficulty in influencing public discourse and political processes. Specifically, civil society organizations such as welfare groups, educational groups, and civic organizations have pointed out that they have fewer opportunities to make direct contact with political actors. However, social media has equipped civil society organizations and stakeholders with new tools that allow them to effectively share information and communicate with the government and to distribute information about their specific interests and missions. Our chapter attempts to discover the ways in which social media usage will affect the behavioral patterns of civil society organizations. The following two research questions were established to investigate any divergence in the behavioral patterns of the civil society organizations.

RQ1: Who are the leading actors who use social media such as Facebook for interaction with local governments?

RQ2: Are Japanese civil society organizations effective advocates when using Facebook?

Methodology

In this chapter, we focus again on the Japanese Facebook community page—the "Tsukuba Civic Activities Cyber-Square,"—which aimed to enhance civil society activities in Japan. Our rationale for analyzing this specific page is based on the findings of the *Chiiki SNS Kenkyūkai* of March 2013. An analysis of the activity levels of the 466 Facebook pages analyzed in 2013 indicate how the "Tsukuba Civic Activities Cyber-Square" can be considered to be a representative case for examining civic engagement via social media in Japan. Detailed frequency regarding engagement and number of fans can be confirmed from Kaigo and Okura (2016).

Facebook metrics data of the "Tsukuba Civic Activities Cyber-Square" was obtained through the Insight interface of Facebook community page functions. Node XL was employed to measure network statistics based on the data derived from the Facebook user accounts. In addition, qualitative content analysis software, KH Coder, was also used in this chapter to identify policy-related articles in the community page content and discourse in Japanese. Our rationale for utilizing this data is the assumption that there are direct and indirect connections between non-profits and governments enhancing policy-making processes.

Users of the Tsukuba Civic Activities Cyber-Square

As we have discussed in previous chapters, the Tsukuba Civic Activities Cyber-Square Facebook Page was created on February 1, 2012, to activate social networks within the Tsukuba community and enhance the level of civic activities (Appendix C). By investigating the residential profiles of the users, we have found that more than 50 percent of users are inhabitants or have their main social activity in Tsukuba, Ibaraki or other cities in the vicinity. This data indicates that the Facebook page is basically used by local individuals and groups.

In Chap. 5, we observed how Tsukuba City had been testing how advertisements and offline gatherings could lead to the further enhancement of civil society through the Tsukuba Civic Activities Cyber-Square. In our qualitative analysis of Tsukuba Civic Activities Cyber-Square posts, we identified how interviews among civil society organizations in Tsukuba have been conducted and their opinions have been featured on the Tsukuba Civic Activities Cyber-Square in news article formats with photographs. Various organizations which focus on environmental protection, welfare, education, and neighborhood associations that operate on maintaining communities in Japan have been featured in the Tsukuba Civic Activities Cyber-Square. This chapter analyzes how these policy-related articles and their reactions via social networking services reflect the actual network and advocacy activities through the Tsukuba Civic Activities Cyber-Square.

For this chapter, we used social network analysis to identify vital actors who exist in the social network community of Tsukuba Civic Activities Cyber-Square. The indicators of: (1) closeness centrality; and (2) clustering coefficient can be used to measure the level of how actors are involved with each other in a network.

Closeness centrality indicates the patterns of how actors make ties. These centrality measures can be used to see the ties among those in non-profit advocacy networks (Borgatti 2005). According to Borgatti, the closeness centrality can be used to measure the degree of how much an actor directly communicates with other actors. The visualization of advocacy networks allows us to see the features and other patterns of the different organizations that are involved with the Tsukuba Civic Activities Cyber-Square Facebook page. Node XL can analyze these network indicators and identify these subgroups in non-profit advocacy networks (Smith et al. 2009). Further analysis of the cluster can identify the subgroups and this sort of analysis can allow us to understand the networking of subgroups and view how they may be leading non-profit advocacy (Hansen et al. 2011). The results illustrate how advocacy activities are supported by actors in a network, allowing us to focus on who is leading each subgroup. These results will provide an insight into the patterns of advocacy activities in a subgroup by the leading actor through the centrality indices.

We chose five different policy cases: (1) the neighborhood association networks (specifically those located in Tsukuba), (2) the educational networks, (3) the disaster networks (especially focusing on the networks during the Hojō tornado in May 2012), (4) the social welfare networks, and (5) the volunteer networks. Previous studies pointed out that the neighborhood association and educational organizations have greater direct access to political elites such as political parties, bureaucrats, local governments, and so on, while social welfare groups (including groups involved with volunteering and disaster reconstruction activities) have less direct access to the political elites. In comparison, we keep in mind these two different types of network structure; we will also illustrate how these different groups have different structures.

Results

Through the available data based on various interview articles about civil society organizations and the reactions posted through the Tsukuba Civic Activities Cyber-Square Facebook page, the result of this analysis identified five policy-related articles and analyzed the interactions of (1) neighborhood association networks (specifically those located in Tsukuba), (2) educational networks, (3) disaster networks (during the Hojō tornado), (4) social welfare networks, and (5) volunteer networks.

Neighborhood Association (NHA) Networks

Figure 6.1 describes the network among the actors who interacted with the neighborhood association-related article. The cluster (the dots in the black circle) in the center of Fig. 6.1 consists of: (1) heavy users of the Tsukuba Civic Activities Cyber-Square who frequently interact with the articles; (2) Tsukuba municipal hall workers; and (3) representatives of the neighborhood associations. These actors are indicated by the black circle. (These were the users who have clicked "Like" more than once on the Tsukuba Civic Activities Cyber-Square, as was reported by a part-time worker in Tsukuba municipal hall responsible for interviewing the Tsukuba Civic Activities Cyber-Square during the Fiscal Years 2013 and 2014.)

These three categories of actors display high centrality, and individuals of Tsukuba City are connected in the vicinity of those who are in these three categories. Figure 6.1 shows the linking among the representatives of the neighborhood associations and Tsukuba municipal hall workers in the network.

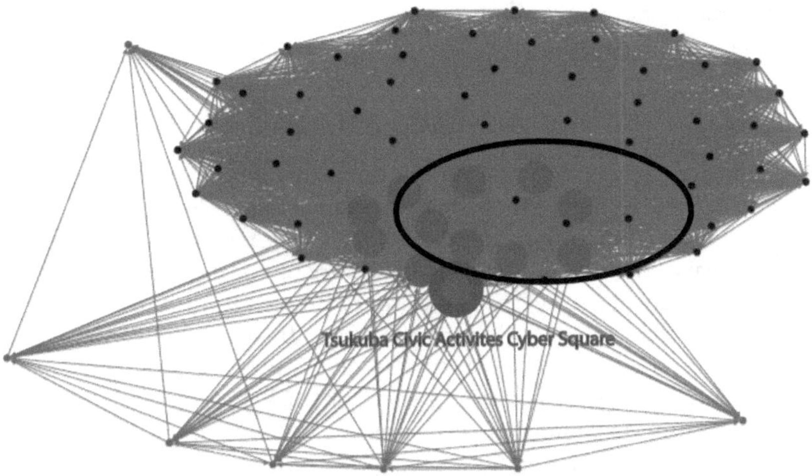

Fig. 6.1 Node XL analysis result map of neighborhood associations (June 12, 2015)

Through our qualitative analysis, we discovered a noteworthy comment by the neighborhood association representative, stating how important it is to be connected in a horizontal relationship and that it is important for the city's government to explain what the neighborhood associations are to other members in the community. Specifically, one representative of a neighborhood association posted the following comment: "I heard that municipal government would be holding a meeting with the social welfare council, neighborhood association, and social workers. Very Good. A horizontal network is important. Please publicize the fact that the ward assemblies (*Kukai*) and neighborhood associations (*Jichikai*) are the same organizations." Other neighborhood association participants also posted comments such as "Recently, communities with children and families are growing rapidly in Tsukuba. I would like to make a 'home' where the children can return to when they become adults." The municipal hall workers responded with gratitude to this being pointed out to them, and they promised that they would endeavor to spread this information. In another case, a representative of a neighborhood association posted a comment on a newsfeed regarding the small festival held in Kukizaki district of Tsukuba City stating, "It is a wonderful festival! I did not know about it. Because I had never participated in the festival, I felt that Kukizaki district was a different municipality" (November 6, 2016), and the other representative of a neighborhood association responded, "Please come and join us!" to the post as well as to the municipal hall workers.

Educational Networks

In an analysis of the network among the actors who interacted with the articles related to topics about education in January 2017, we observed how the links among users of the Tsukuba Civic Activities Cyber-Square, educational organizations, and other users are of equal distance within the network. There were no specific posts or conversations, and they connected via "sharing" and "liking" the posts. In other cases, however, a member of an educational civil society group made a comment on the newsfeed regarding the scientific events for small children, stating: "It is very interesting for adults to have scientific experiments with small children. It is also wonderful that we can exchange communication about volunteering. Children can have great experiences that will enhance their development with their parents. Let us expand the scientific culture in Tsukuba!" (September 14, 2016).

Disaster Networks

Figure 6.2 describes the network among the actors who interacted with the articles related to the tornado disaster of Tsukuba City's Hōjō area in May 2012. The cluster of actors in the circle 1 are the Tsukuba municipal hall workers and heavy users linked to this group of actors. The cluster of actors in circle 2 are welfare workers and heavy users who are linked to this group of actors. The cluster of actors in the circle 3 are actors in education and research and heavy users who are linked to this group of actors.

In another case, one of the citizens in Tsukuba posted a comment to a newsfeed regarding the reconstruction activities conducted by one of the social welfare groups from the Great East Japan Earthquake telling municipal hall workers, "I support your activities."

Social Welfare Networks

In analysis of the network among the actors who interacted with the articles related to welfare and medical services we observed how the Tsukuba Civic Activities Cyber-Square linked to the welfare-related organization involved in rescue education is linking other users of the Tsukuba Civic

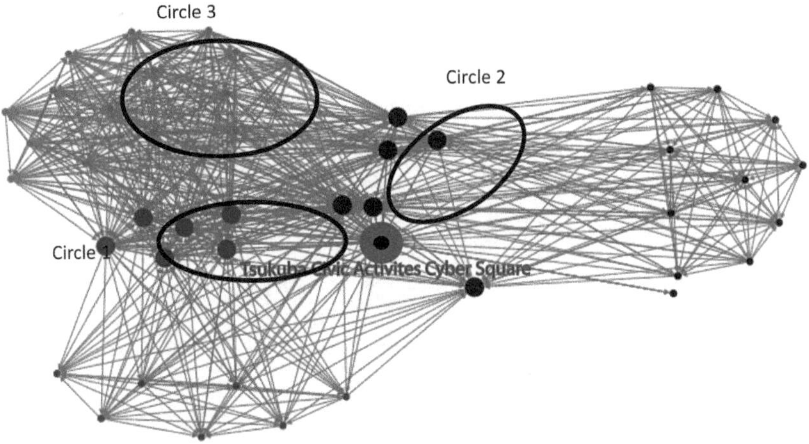

Fig. 6.2 Node XL analysis result map of Tornado in Hōjō, Ibaraki (June 12, 2012)

Activities Cyber-Square community. Through our qualitative analysis, we found that the welfare group expressed their gratitude to the daily support of municipal hall workers, and the workers also expressed appreciation for their words and encouraged the group members to be more active. Specifically, the welfare group commented "I truly appreciate your kind support to our activities," to which the municipal hall worker responded "I should say thanks to you likewise. Looking forward to your success in the next academic year." In other cases, a member of a welfare group commented on the newsfeed, "Please pay particular attention to community-based health care even if you are currently healthy, and take part in our team as soon as possible" (March 5, 2016), and encouraged the municipal government to have more interest in the welfare problem.

Volunteer Networks

Figure 6.3 describes the network among the actors who interacted with the articles related to volunteering. Through this network map, one can observe how the links among users of the Tsukuba Civic Activities Cyber-Square, welfare organizations, and leisure-related organizations are of equal distance within the network. In contrast to the other networks which had a central actor, this represents how many of the actors are linked equally to each other.

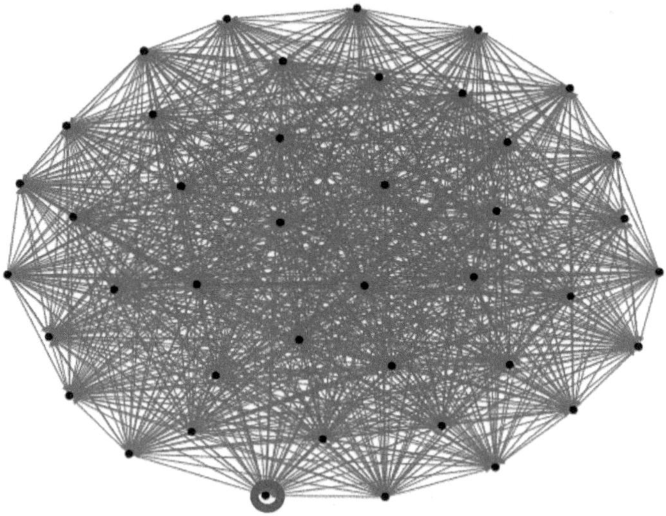

Fig. 6.3 Node XL analysis result map of volunteering (January 28, 2016)

Through our qualitative analysis, we found that volunteering groups expressed their gratitude and opinions through the newsfeed. For example, on September 15, 2016, one volunteering group commented, “I would like to change information regarding volunteering. Children can join scientific activities just like adults. Let’s expand a grassroots network throughout Tsukuba!” The group encouraged the participants of the Facebook page and the municipal hall workers to engage more in this specific type of volunteering.

Analysis Results and Comparison

Our analysis of network shapes among the organizations links can be generalized as follows (Table 6.1) and our main findings are as follows.

First, we could confirm that the disaster network structure consisted of welfare groups, civil servants in Tsukuba City, and other entities. Previous studies have indicated that these are people in groups that have fewer chances to advocate to political elites directly, but they were found to have a similar network structure to the neighborhood associations in our study results. This is noteworthy because previous studies have pointed out that

Table 6.1 Comparison

	The center (leading actors)	*The periphery*
NHA network	NHA Civil servants in Tsukuba Users of the Tsukuba Civic Activities Cyber-Square	Users of the Tsukuba Civic Activities Cyber-Square
Educational network	None	None
Disaster network	Welfare groups Civil servants in Tsukuba Users of the Tsukuba Civic Activities Cyber-Square	Academic/educational groups Users of the Tsukuba Civic Activities Cyber-Square
Welfare network	Tsukuba Civic Activities Cyber-Square	Welfare groups Users of the Tsukuba Civic Activities Cyber-Square
Volunteering network	None	None

neighborhood associations have more opportunities to connect to political elites such as those in local governments. Specifically, the neighborhood association representatives in our analysis were found not only to be hubs, but were also communicating requests and demands to be delivered to the local government through Facebook and were able to obtain a positive response from the local government that they would sincerely consider their request. We could confirm that neighborhood associations had an important function to advocate to the local government as previous studies had already pointed out, and that they are among the traditional advocacy routes, especially at the local level in Japan. At the same time, interestingly, when we look at disasters, the municipal hall workers and welfare organizations are creating a hub within the network. When it pertains to neighborhood associations (community maintenance and community building groups), representatives of those associations, municipal hall workers and other users of the community page are at the center of the network (Appendix D). These results indicate that smaller organizations and citizens have more opportunities to connect to the local government via Facebook by becoming hubs of the information and advocacy activities.

Second, with regard to the welfare network, the welfare groups were able to have some online conversations with the local government, and advocated their political concerns to them, just as the neighborhood association representative did in the result of our analysis of the neighborhood association network. Specifically, one member of a welfare group made a comment on the newsfeed that they would be able to gain more attention and as a result, possibly influence the welfare policies through the municipal hall workers. In the same way, the neighborhood association representative posted a comment on the newsfeed in order to establish a horizontal network between local government and citizens. This result indicates that platforms such as Facebook may provide more opportunities for newer and smaller groups with limited resources to advocate their concerns to local governments as well as connect with them directly.

Third, in a social networking service community page focusing on civil society activities, networks among actors and organizations do not need a hub when it concerns volunteer activities and social welfare activities. Similar to volunteer activities and social welfare activities, educational activities do not have a hub within the network, and actors or organizations can connect directly to each other.

These results indicate how social media can be a platform to provide opportunities for direct communication with local governments and advance advocacy activities. This becomes vital especially for social welfare groups that traditionally do not have many opportunities to have direct communication with local governments in Japan and where such civil society organizations are lacking in employees, when compared with other nations. In other words, a platform like Facebook could (1) provide more chances to connect with the local government, (2) provide more political opportunities to advocate, and (3) create opportunities to exert greater presence.

Better Advocacy Through Social Media

Our analysis discovered that diverse patterns of usage were evident in nonprofit advocacy among the leading actors connected to the Tsukuba Civic Activities Cyber-Square. In this section, we will conclude this chapter with our main findings.

First, we could confirm that the disaster network structure in the result of our analysis consisted of welfare groups, civil servants in Tsukuba City and other entities. Previous studies had pointed out that these people have fewer opportunities to advocate to political elites directly; however, we found that they have a similar network structure as the neighborhood associations, which have more chances to connect to the political elite. Both welfare groups and neighborhood association representatives in our analysis were found to be hubs in the communication flow as well as when communicating requests and demands to be delivered to the local government through Facebook. These results indicated that smaller and newer organizations and citizens have more opportunities to connect with local governments via Facebook by becoming the hubs of the information and advocacy activities.

Second, with regard to the welfare network, the welfare groups were able to have some online conversations with local governments, and advocated their political concerns to them, in the same manner as the neighborhood association representative, through the results of the analysis of the neighborhood association network. A member of a welfare group made a comment through Facebook in order to get more attention to the welfare policies from the municipal hall workers. We suggest that the dynamic that we observed indicates that social media platforms such as Facebook may provide more opportunities for newer and smaller groups with limited resources. These smaller entities can possibly have a better chance to advocate to local governments as well as connect to them directly.

Third, in a social networking service community page focusing on civil society activities, networks among actors/organizations do not need a hub when it concerns volunteer activities and social welfare activities. Similar to the situation with volunteer activities and social welfare activities, educational activities do not have a hub within the network, and actors and organizations could connect horizontally with each other.

These results show that civil society organizations such as welfare groups that were once categorized as "silent" groups, displayed equal amounts of advocacy in comparison to other educational organizations and traditional communities such as neighborhood associations. The results of the qualitative content analysis suggest that these organizations are not just attempting to exchange and share information with the local government via Facebook, but are also able to advocate their concerns and issues they would like to pursue. To illustrate this dynamic, a member of a welfare group made a comment on the newsfeed to encourage the municipal government to show a greater interest in welfare issues. In another case, one volunteering group encouraged the participants on the Facebook page and the municipal hall workers to engage more in volunteering.

Our findings indicate that social networking services such as Facebook could provide civil society organizations with (1) more chances to connect with the local government and (2) more political opportunities to advocate and could (3) create opportunities to exert greater presence, despite their limited financial and political resources.

References

Bass, G. D., Arons, F. D., Guinane, K., & Carter, F. M. (2007). *Seen but not heard: Strengthen nonprofit advocacy.* Washington, DC: Aspen Institute.

Bass, G. D., Abramson, J. A., & Dewey, E. (2014). Effective advocacy: Lessons for nonprofit leaders from research and practice. In R. Pekkanen, R. S. Smith, & Y. Tsujinaka (Eds.), *Nonprofits and advocacy: Engaging community and government in an era of retrenchment* (pp. 254–293). Baltimore, MD: Johns Hopkins University Press.

Borgatti, S. (2005). Centrality and network flow. *Social Network, 27*(*1*), 55–71.

Chiiki SNS Kenkyūkai. (2013, March). *List of Facebook pages run by local governments in Japan.* Retrieved from http://www.local-socio.net/localgovernment_facebookpage_20130323.pdf

Child, C. D., & Grønbjerg, A. K. (2007). Nonprofit advocacy organizations: Their characteristics and activities. *Social Science Quarterly, 88*(1), 259–281.

Dahlberg, L. (2001). The Internet and democratic discourse: Exploring the prospects of online deliberative forums extending the public sphere. *Information, Communication & Society, 4*(4), 615–633.

Hansen, D., Shneiderman, B., & Smith, A. M. (2011). *Analyzing social media networks with NodeXL: Insights from a connected world.* Burlington, VT: Morgan Kaufmann Publishers.

Kabashima, I. (1990). Masu media to seiji [Mass media and politics]. *Leviathan, 7*, 7–29.

Kabashima, I. (2004). *Sengo Seiji no Kiseki: Jiminto shisutemu no kiseki to henyō* [*Trajectory of postwar Japan: Formation and transformation of the LDP system*]. Tokyo: Iwanami Shoten.

Kabashima, I., & Broadbent, J. (1986). Referent pluralism: Mass media and politics in Japan. *Journal of Japanese Studies, 12*, 329–361.

Kabashima, I., Takeshita, T., & Serikawa, Y. (2007). *Media to Seiji* [*Media and politics*]. Tokyo: Yuhikaku.

Kaigo, M., & Okura, S. (2016). Exploring functions in citizen engagement on a local government Facebook page in Japan. *Telematics & Informatics, 33*(2), 584–595.

Kavanaugh, A. L., Fox, E. A., Sheetz, S. D., Yang, S., Li, L. T., Shoemaker, D. J., Natsev, A., & Xie, L. (2012). Social media use by government: From the routine to the critical. *Government Information Quarterly, 29*(4), 480–491.

Keane, J. (2000). Structural transformation of the public sphere. In K. L. Hacker & J. A. G. M. Van Dijk (Eds.), *Digital democracy: Issues of theory and practice* (pp. 70–89). London: Sage Publications.

Kobayashi, T. (2013). Masumedhia yorimo "chūritsu"na nihon no netto nyūsu: 2012 nen shūgiin senkyoji no nyūsu riyou ni okeru tōhasei baiasu ninchi [Internet news which is more moderate than mass meidia in Japan]. In S. Kiyohara &

K. Maejima (Eds.), *Netto senkyo ga kaeru seiji to shakai* [*Effect of the Internet on politics and society*] (pp.119–147). Tokyo: Keio University Press.

Lee, G., & Kwak, Y. H. (2012). An open government maturity model for social media-based public engagement. *Government Information Quarterly, 29*(2), 492–503.

Leech, B. (2010). Lobbying and influence. In L. S. Maisel & J. M. Berry (Eds.), *The Oxford handbook of American political parties and interest groups*. Oxford: Oxford University Press.

Leng, R. (2015). Japan's civil society from Kobe to Tohoku: Impact of policy changes on Government-NGO relationship and effectiveness of post-disaster relief. *Electron Journal of Contemporary Japanese Studies, 15*. Retrieved from http://japanesestudies.org.uk/ejcjs/vol15/iss1/leng.html.

Linders, D. (2012). From e-government to we-government: Defining a typology for citizen coproduction in the age of social media. *Government Information Quarterly, 29*(4), 446–454.

Ogasawara, M. (2014). Sōsharu medhia jou no seiji comyunikēshon to masumedhia [Mass media and political communication via social media in Japan]. *Journal of Mass Communication Studies, 8*, 63–80.

Okamoto, T., Ishibashi, S., & Wakisaka, T. (2015). Netto senyko kaikin no kouka wo kenshō suru: 2013 nen saninsen deno tōhyō ishi kettei ni taisuru kouka no bunseki [Measuring the impact of the Internet campaigning liberalization on voters: An analysis using the data from the 2013 upper house election of Japan]. *The Law Review of Kansai University, 64*(6), 1879–1900.

Okura, S. (2013). Nihon no fukushi bunya ni okeru dantai doukou no fukusen sei: Dainiji dantai no kiso kouzou ni kansuru chōsa wo mochiite [Double track change of social welfare organizations in Japan: Evidence from "J-JIGS2"]. *Area Studies Tsukuba, 34*, 271–294.

Pekkanen, R. (2006). *Japan's dual civil society: Members without advocates*. Stanford: Stanford University Press.

Pekkanen, R., & Smith, R. S. (2014). Nonprofit advocacy in Seattle and Washington, DC. In R. Pekkanen, R. S. Smith, & Y. Tsujinaka (Eds.), *Nonprofits and advocacy: Engaging community and government in an era of retrenchment* (pp. 47–65). Baltimore: Johns Hopkins University Press.

Salamon, L. M., & Geller, L. S. (2008). Nonprofit America: A force for democracy? *Communiqué, 9*, 1–25.

Sassi, S. (2000). The controversies of the Internet and the revitalization of local political life. In K. L. Hacker & J. A. G. M. Van Dijk (Eds.), *Digital democracy: Issues of theory and practice* (pp. 90–104). London: Sage Publications.

Schwartz, F. (2003). What is civil society? In J. F. Schwartz & S. J. Pharr (Eds.), *The state of civil society in Japan*. Cambridge: Cambridge University Press.

Smith, M. A., Shneiderman, B., Milic-Frayling, N., Rodrigues, E. M., Barash, V., Dunne, C., Capone, T., Perer, A. & Gleave, E. (2009). Analyzing (social media) networks with NodeXL. In J. M. Carroll (Ed.), *Proceedings of the fourth international conference on communities and technologies*, C{&}T 2009, University Park, PA, USA, June 25–27, 2009 (pp. 255–264). ACM. http://doi.org/10.1145/1556460.1556497

Tsuda, D. (2012). *Uebu de seiji wo ugokasu!* [*The Internet can change the politics!*]. Tokyo: Asahi Shimbun Publications.

Tsujinaka, Y., & Ito, S. (2010). *Rōkaru gabanansu: Chihō seifu to shimin shakai* [*Local governance: Local governments and civil society*]. Tokyo: Bokutakusha.

Tsujinaka, Y., & Mori, H. (2010). *Gendai shakai shūdan no seiji kinō: Rieki dantai to shimin shakai* [*Political functions of social organizations in contemporary Japan: Faces of interest group and those of civil society*]. Tokyo: Bokutakusha.

Tsujinaka, Y., Choe, J., & Kubo, Y. (2010). Nihon no dantai bunpu to risōsu: Kokka hikaku to kokunai chiiki kan hikaku kara [The group distribution and its resources]. In Y. Tsujinaka & H. Mori (Eds.), *Gendai shakai shūdan no seiji kinō: Rieki dantai to shimin shakai* [*The political function of civil society organization in contemporary Japan*] (pp. 33–64). Tokyo: Bokutakusha.

Wilhelm, A. G. (2000). *Democracy in the digital age: Challenges to political life in cyberspace*. New York: Routledge.

Yamamoto, H. (2010). Rieki dantai no eikyōryoku: Takaku teki na shiten kara miru kenryoku kouzou [The influence of interest groups: Power structures from various perspectives]. In Y. Tsujinaka & H. Mori (Eds.), *Gendai shakai shūdan no seiji kinō: Rieki dantai to shimin shakai* [*Political functions of social organizations in contemporary Japan: Faces of interest group and those of civil society*] (pp. 37–252). Tokyo: Bokutakusha.

Yamazaki, A. (2015). Intānetto senkyo undou to yūkensha no jōhō sesshoku: 2013nen sangiin senkyo to 2012nen shūgiin senkyo no hikaku [Online campaign and information behavior among Japanese voters: Comparing the 2013 upper house election with the 2012 lower house election]. *Senkyo kenkyū* [*Annals of the Japan Election Studies Association*], *31*, 102–113.

Sae Okura is an assistant professor in the Faculty of Humanities, Law and Econoimcs at Mie University in Japan. She conducts research in politics and social policy studies. Her recent research interests focus on political participation by minority groups such as the disability community, sexual minorities, and single parent families. She analyzes quantitative survey data collected in Japan.

Muneo Kaigo is an associate professor in the Faculty of Humanities and Social Sciences at the University of Tsukuba in Japan. He teaches courses in media communication in the Graduate School of Humanities and Social Sciences, media management in the Graduate School of Business Sciences and communication science in the College of Comparative Culture at the University of Tsukuba. He is currently leading a joint research project on social media uses among local municipalities in Japan with cooperation from the municipal government of Tsukuba and Intel Corporation Japan.

CHAPTER 7

Conclusion

Muneo Kaigo and Sae Okura

Shaken into the Age of Social Media?

Some time has now passed since the Great East Japan Earthquake. Often, the memories of all catastrophes and consequences gradually fade away. The Great East Japan Earthquake is now no longer the most important topic for many Japanese. Many of the images of the Great East Japan Earthquake are still able to trigger memories of the Japanese, as the visuals of destruction are revisited by the media every March 11, the anniversary of the earthquake. The problems around the Fukushima Daiichi nuclear power plant meltdown recovery effort will continue for decades, and each report of progress will remind the Japanese of the catastrophic events that occurred in 2011. The trauma among many of the citizens in the worst-affected areas may never heal. Taken as a whole, it is clear that the earthquake has altered both the country and its people in many ways since March 11, 2011.

M. Kaigo (✉)
Faculty of Humanities and Social Sciences, University of Tsukuba in Japan, Tsukuba, Ibaraki, Japan

S. Okura
Faculty of Humanities, Law and Economics, Mie University, Tsu, Mie, Japan

M. Kaigo (ed.), *Social Media and Civil Society in Japan*,
DOI 10.1007/978-981-10-5095-4_7

The negative consequences that still emanate from this multi-stage catastrophe are being researched in many fields in Japan and it is to be hoped that the results will better prepare Japan for the next large-scale disaster. During the time this tragic disaster occurred, the attitudes of Japanese towards media were gradually changing. This coincidence in timing may have resulted in the disaster acting as a catalyst in quickening this change of attitude towards the media in Japan. The first chapter of this book discussed how wary the Japanese had been about using the Internet in the past, and we have observed in the rest of the study just how the Japanese have shifted quickly into becoming increasingly reliant on the Internet.

The first chapter covered the Japanese Internet environment and focused on the differences between computer anxiety and the cultural construct of the Japanese *nigate-ishiki* or low self-efficacy of information and communication technology (ICT). For some time, the sluggishness over the adoption of computers in Japan limited the total range of information available for many Japanese. This was the result of many Japanese only using devices with limited access, such as web-enabled mobile phones. Many were relying on access to the Internet solely through mobile phones and the Japanese Internet environment evolved to be dependent on the small screen of these web-enabled mobile phones.

The inconvenience of substituting a telephone keypad is cumbersome for those who are accustomed to the QWERTY keyboard; however, in the case of the Japanese and also for many nations with languages that are non-alphabetic, some sort of character substitution is required for input processing. In Japan, the telephone keypad became one of the popular and preferred input methods, and it allowed many Japanese to finally adopt ICT usage and the Internet. High-speed mobile data access evolved quicker in Japan because of the dependence on the mobile phone and this led to a unique and peculiar evolution of mobile phone development among Japanese electronics manufacturers.

Chapter 1 also introduced several prominent, but negative facets of the Japanese Internet environment, including the *Ni-Channel* and *Matome Saito*, which have also led the public to be wary of the Internet. The *Ni-Channel* and *Matome Saito* are representative platforms that highlight the dark and sinister nature of Internet users in Japan. Even worse, the highly prominent viral infection of Japanese computers through the Winny peer-to-peer file sharing platform scared away many Japanese who were ambivalent regarding accessing the Internet through computers. Even the wide variety of content available on the Internet led people to think that it lacked the wholesomeness to be allowed in the home. The regulatory standards used for evaluating the decency of content of commercial mainstream media left the Japanese shocked when they saw that horrifying images of death or obscenity were readily available through the Internet.

Through this unintentional combination of chance, the Japanese were creating an impression of the Internet as a dark and scary environment. Accordingly, the Japanese were very cautious with regard to adopting ICT, especially Internet-related devices. The impression was given that computers should be handled by the experts because amateurs will soon end up having their computers infected and distributing sensitive personal information to criminals. It was thought that children would be able to access illegal pornography or end up dead after accessing suicide assistance bulletin boards. To most Japanese, the Internet has been perceived as an environment that was full of danger and therefore in stark contrast to the mostly safe society of Japan. The Internet was perceived to be something that was to be guarded against and that parents needed to limit the exposure among children and keep them in the safe, real world.

One commonly known term has often been used to describe this trait of the Japanese, *Galapagos*. The term or metaphor derives the perception that the Japanese are so good at isolating themselves as they are the residents of an archipelago of islands. The somewhat self-deprecating usage of the metaphor of *Galapagos* is often also used to describe the unique evolution of the Japanese with regard to electronics, computers, and mobile ICT. The most common item of this peculiarly national phenomenon were the highly advanced, web-enabled multifunctional mobile foldable phones that dominated Japan during the 2000s, and these have been commonly known by the abbreviated term, *gara-kei* (*garapagosu keitai* = Galapagos-style mobile phones) among Japanese. The *gara-kei* was the primary device for the development of ICT in Japan. The Japanese ICT environment and even social media, for a time, was proceeding with its own style of evolution, until smartphones arrived and then the Great East Japan Earthquake.

The ability to be mobile and to be less reliant on an electrical outlet were two things that Japanese became increasingly aware of during and after the Great East Japan Earthquake. Along with television sets and stereo systems, computers normally require constant or frequent connection to an electrical outlet. The batteries of today's laptop computers and tablets do not require frequent charging; however, at the time of the Great East Japan Earthquake, laptop computers did not generally have such long-lasting batteries and therefore quickly required recharging. As discussed earlier, during the Great East Japan Earthquake, a large area of Japan lost its electricity supply. In addition, in the days following the debacle of the Fukushima Daiichi nuclear power plant meltdown, many areas of Japan had to start preparing for planned power outages. These planned power outages were already taking place in Tokyo immediately after the earthquake as a preemptive measure designed to avoid a widespread blackout because of the lack of electricity. Following the earthquake, all of the country's nuclear power plants had to be shut down because of the concerns following the Fukushima Daiichi accident. *Keikaku-teiden* or planned power outages were for a long time unheard of in Japan before the Great East Japan Earthquake. The expression was last heard during days following the post-World War II defeat and the ensuing chaos. It was clear that during and after the Great East Japan Earthquake Japan was experiencing another catastrophe similar to the post-World War II-style crisis situation.

In result, the planned power outages following the Great East Japan Earthquake acted as catalysts for the Japanese to adopt devices and technology that are not reliant on having constant electricity procurement through outlets. Companies adopted standby supply systems and smartphones and tablets were being adopted as perfect auxiliary devices for accessing the Internet in the event of planned power outages. Wireless connectivity with devices that had computer-like Internet access and less dependency on having an electrical outlet were catalysts that pushed the Japanese out of their *Galapagos* ICT environment.

The Fukushima Daiichi nuclear power plant accident also played a role as a catalyst for changing the general attitude toward media, especially mainstream media, among the Japanese. As the meltdown and explosions of the power plants were being televised and reported, the mainstream media were inadvertently leaving the public in the dark about what was actually happening. The media themselves were not receiving any vital information, and, ironically, this had the consequence of gradually lowering the credibility of the mainstream media. The Tokyo Electric Power Company did not immediately disclose information of what was happening for the media as a result of confusion or incompetence, and the Japanese cabinet at the time was ill-prepared to manage such a major nuclear power plant accident. To make matters worse, the Japanese prime minister and cabinet at that time had a poor relationship with the Japanese ministries and high-level bureaucrats. Instead of letting the administrative vice-ministers or undersecretaries have increased control of the ministries as had traditionally had been the case, the newly appointed ministers themselves had been claiming accountability by adopting a new policy to better control them and taking control over press conferences. In this crisis situation, the nuclear power plant disaster left the government in disarray along with the Tokyo Electric Power Company, the owner of the Fukushima Daiichi nuclear power plant. As the meltdown of the plants continued, the Japanese public were making extensive efforts to find more information to better understand and assess the situation, but the mainstream media were unable to provide any worthwhile information about what was happening, or would happen, in Fukushima. During this period, social media became the new channel of information providing various types of information from different sources, not all of them accurate, but still worthwhile reading or viewing and information pertaining to the needs of the public. Social media was providing news, answers and expert information in relation to the following questions that the mainstream media were not providing. What is happening at the Fukushima Daiichi nuclear power plant? What will happen if there is a meltdown? Who will be affected? What precautions should we take? What will we need? The media was falling behind, as people were increasingly seeking for information through social media.

In a sense, the level of mainstream media dependency of the Japanese effectively changed after the Fukushima Daiichi nuclear power plant accident. Media dependency on the mainstream media among Japanese has since been lower among many younger Japanese as the information being provided by the mainstream media was both limited and slow, especially with regard to accessing knowledge about the accident. The public increased its demand for more timely information and an increase in the news about what was actually happening. The usual method of providing information through telecasts and newspapers or websites was too slow and the format demanded too much time of the audiences. Users were realizing that glancing at the tweets and retweets of the same information is instantaneous, requiring only seconds. Other social media were able to provide detailed information by experts on speculation of what was going to happen to the Fukushima Daiichi nuclear power plant, while the mainstream media were still trying to verify their facts and sources. The combination of smartphones and social media such as Twitter and Facebook which give easy access to information from overseas also shifted many Japanese into the age of social media usage. The Japanese skepticism towards the mainstream media has fostered the strong demand for alternative information sources, and social media was the perfect fit at the time.

At the local level, the Great East Japan Earthquake also shifted the Japanese into social media usage, albeit in a different context. As documented in the second chapter, Twitter was used at the local government level to send and share information immediately after the Great East Japan Earthquake as occurred, for example, in the case of Tsukuba and other municipalities. Social support and trust was formed and became an important source for gaining vital information in the days after the large-scale disaster. The relatively high pre-existing level of social capital among Japanese was advantageous and the characteristics of Twitter were, accordingly, beneficial for many citizens of Tsukuba and other areas with mid-scale disruptions. The success of Twitter usage led to the subsequent launch of the Tsukuba Civic Activities Cyber-Square of Tsukuba City, and the diffusion of other social media such as Facebook has reduced online anonymity (Park et al. 2011). This enables local governments in Japan to be able to communicate with their citizens with less concern of the issues that Japanese have towards the Internet. Chapter 4 also illustrates how the Tsukuba Facebook page, the Tsukuba Civic Activities Cyber-Square functioned during another subsequent disaster, a destructive tornado. Information about damage caused by the tornado came in quickly through Facebook, and within one week, users began providing information to support the disaster victims. Information about volunteering was distributed and this actively assisted the relief efforts. Social media was found to be useful in both instances of Tsukuba related disasters. The findings here can potentially be used as a guide on how to navigate social media during these situations. In a sense, the Great East Japan Earthquake made everyone a stakeholder for keeping the community together. Social media users and the government worked together to share information and help each other during the disasters (Fig. 7.1).

Since the Great East Japan Earthquake, Japan has been avid in disseminating information about disaster preparation, as municipal governments across Japan are preparing prevention maps. Facebook has a safety check-in service, and all telecommunication carriers have voicemail services such as message dials or message boards to so that people can check the safety of others during congestion. Japanese apps for smartphones provide early earthquake warning systems and safety tips are circulated periodically meaning that people will be aware of these important resources. Another significant change since the Great East Japan Earthquake is that many of these resources are now provided in English or have English-language instructions on how to use these services.

Most local governments in Japan have already initiated some sort of use of social media to cope with crises and disasters, as a significant portion of the population has access online. The problem lies in the management of

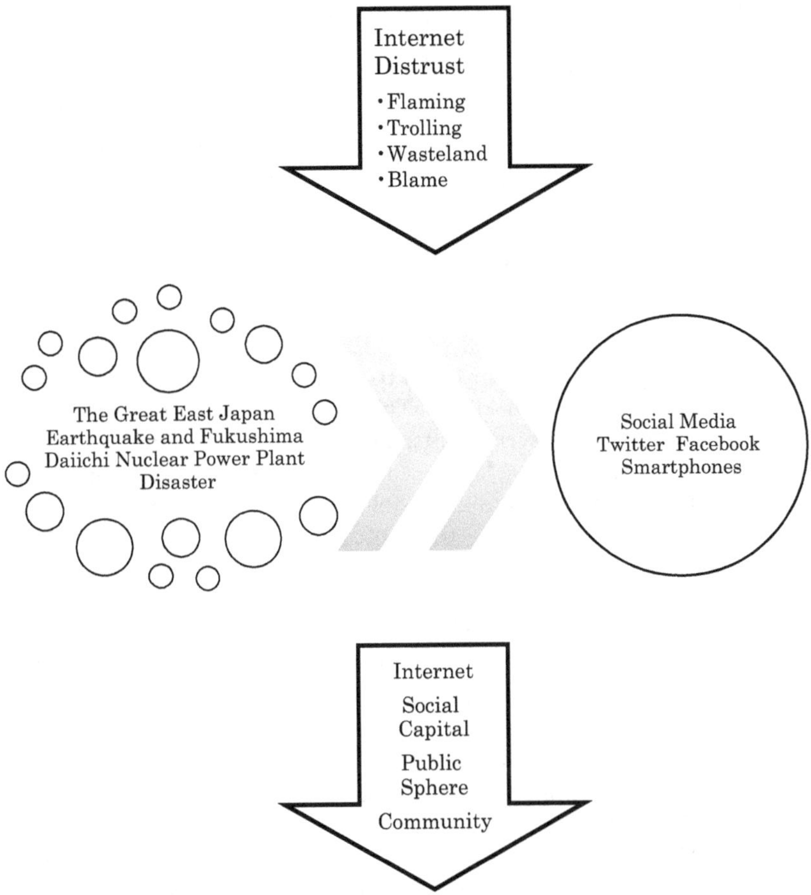

Fig. 7.1 The transformation of the Japanese perception of the Internet

information and crisis management planning by the local governments (Wendling et al. 2013; Freberg 2012). Social media is an interactive platform and is quite different from the traditional practices of strategic communications and public relations. The citizens also have a voice that allows to be heard through the same platforms on level terms. If the local government begins to be perceived as disorganized or lacking interest, the public can undermine any prior crisis contingency plan. In Japan, the neighboring city located next to Tsukuba, Josō City, experienced severe flooding following a prolonged storm in September 2015. As the city hall was located in the affected area, it was also became submerged, and became the victim of *enjyo* on social media as the Josō City Hall was a designated evacuation center. Therefore, although social media networks are effective tools for managing a risk or crisis, the contingency plan itself is of utmost importance or the administration will risk a public relations disaster at critical moments.

The Struggles and Successes with Social Media and Civil Society in Japan

Another main foundation of this book focuses on the struggles and successes by local governments in Japan to adopt social media usage to revitalize communities and civil society. The third chapter examines the policy areas of Japanese local governments, and through an analysis of all of the Facebook pages operated by Japanese local governments, we found a wide variation of policy areas, number of followers and levels of engagement. Although it is difficult to gauge the appropriate number or size of followership for social media to be effective for civil society to communicate, the case of Tsukuba may provide an interesting hypothesis.

Prior to the Great East Japan Earthquake, we know that the number of followers of Tsukuba City Hall's Twitter page had already reached 2000, and this became a solid foundation of social media users that helped the number of followers to grow dramatically during the subsequent disaster. When we examined the Tsukuba Civic Activities Cyber-Square, we were able to observe that during the first and second year the number of followers was under 2000 users, and during this time, the status of the Tsukuba Civic Activities Cyber-Square was akin to being on a life support system. If the community was not being managed, activity within the Tsukuba Civic Activities Cyber-Square would quickly become silent. However, once the Tsukuba Civic Activities Cyber-Square had more than 2000 followers, the community became more sustainable, requiring less maintenance by community managers. Our hypothesis about the sustainability of these sites is therefore related to this number, 2000. We believe that once a community dealing with civil society and local governance reaches 2000 members, the community becomes more self-reliant requiring less intervention by community managers. In other words, 2000 may be the lower threshold for maintaining a virtual community to be self-sustainable.

Our 2000-member hypothesis is also a rejection of the notion that social media allows for anything to become "viral" automatically. We have observed through this project that for something to become viral on social media requires many conditions, and it is not automatic. Especially in the case when people use social media with regard to civil society, a post becoming viral is somewhat rare, and requires a coincidence of many other elements and conditions. On the other hand, in a bad way, viral communication is a common risk that needs to be acknowledged by organizations that use social media. Anything scandalous or prone to criticism is easy prey and can become viral in next to no time. In this sense, viral communication is a possible threat to social media with regard to civil society to which local governments need to pay attention. The Japanese "take" on flaming, *enjyo*, has become even more common in recent years with the advent of social media. In Japan, Twitter is the platform most prone to *enjyo*, as its users often forget about their potential audience—which may be every single person online. The speed of proliferation of information through social media and the potential for comments by others to be uninhibited or extreme, easily lead to this. Once *enjyo* starts, everything is literally up in flames and it is difficult to stop.

Another finding from our project that is related to the field of marketing is how the reach statistic of a posting on Facebook does not always correlate with the value of engagement that is provided through the Facebook Insight interface. The Tsukuba Civic Activities Cyber-Square example has provided evidence that high levels of engagement can coexist with low levels of the reach statistic. What can such a relationship mean? We observed in Chap. 5 how engagement among members of the community would be high on social media even when the level of reach is low. However, we also observed when advertisements that steer in higher levels of reach were initiated, that this would cause engagement among the community to diminish. The results need more verification through support from other research, but our findings suggest how low levels of reach are not necessarily always a bad thing, especially if the issue of engagement is more important for the community. Social media is designed in a way to allow for people to interact and, therefore, the members of the community cannot be passive, but need to be more active. Our hypothesis here is that advertisements may not necessarily be beneficial for social media communities.

One outcome of the Tsukuba Civic Activities Cyber-Square is how social media usage prompted the city hall workers to begin actively collecting information about local neighborhood associations in Tsukuba. Before, the connection between the municipality and the neighborhood associations were more bureaucratic, as documents would be exchanged and reports would be formally processed. Facebook usage helped better facilitate the city hall workers to find and share information about Tsukuba, and brought about the Tsukuba Civic Activities Cyber-Square community managers to visit the neighborhood associations so that they could introduce and share the experiences with other neighborhood associations and citizens. This information dissemination was fundamental in building bridges between the neighborhood associations in different jurisdictions.

One noteworthy example of information dissemination with regard to a neighborhood association in Tsukuba is Shinozaki. Shinozaki has a blend of residents that have lived in the area for generations and those that have just recently moved in. Usually, in many parts of Japan, this type of blend of residents typically results in tensions arising between the newcomers and the old residents, making neighborhood associations difficult to manage. The Shinozaki neighborhood association of Tsukuba is a relatively rare successful anomaly where the blend of new and old residents function in the association harmoniously. To feature their activities, the Tsukuba Civic Activities Cyber-Square interviewed the representatives of the association. This interview later caught the attention of the local newspaper (*Joyō* newspaper) and became a news article. The news article that was partially accessible as a web page then caught the attention of the Fujimi City municipality of Saitama Prefecture, which has itself been struggling to successfully blend new and old residents to function cooperatively in the neighborhood associations. The Fujimi City Hall workers visited Tsukuba City and was provided the information that was originally reported through the Tsukuba Civic Activities Cyber-Square. This type of synergy among local governments, social media and traditional media provides an attractive way of raising interest among civil society. Unfortunately, for the case of Tsukuba, the *Joyō* newspaper will be in hiatus from April, 2017 and will no longer be able to provide information about local neighborhoods in Ibaraki Prefecture.

Regeneration and Cytokinesis of Civil Society Organizations in Tsukuba

After the Great East Japan Earthquake, the residents of Tsukuba experienced heightened levels of social capital and a public sphere that was constructed through social media. Although the natural disaster left the city with considerable damage, it left the city with a heightened expectation of better community building. The Tsukuba Civic Activities Cyber-Square on Facebook was an attempt on the part of the city to expand the network of non-profit organizations and civil society to communicate and cooperate with each other. Social media has been found to be a good vehicle to attain this and Facebook was a good platform to initiate this framework and also a mechanism to permit civil society organizations to begin increased interactions with each other. In the case of the Tsukuba Civic Activities Cyber-Square, the Tsukuba City government became a leader in arranging and getting together the civil society in Tsukuba.

From the viewpoint of the civil society organizations, the online interactions through social media and offline interactions through organized events by the city are both important. Relying solely on social media and interaction through the Internet does not allow for deeper connections or strong networks. However, civil society organizations each have their activities and ideas, so without the Tsukuba Civic Activities Cyber-Square, the organized events would not have been able to gather participants. Through the annual forum to permit civil society groups and organizations to exchange information and share ideas, Tsukuba City has attempted to address some of the concerns expressed by the people in the civil society organizations.

Social media use through the Tsukuba Civic Activities Cyber-Square has played the important role of providing information with value to users that are particularly interested in the topic of civil society. Social media allows the users to engage the content, and this differs highly from the one-way pamphlets and leaflets distributed by the local governments. The Tsukuba Civic Activities Cyber-Square was able to introduce different organizations throughout Tsukuba involved in civil society matters, something that was able to heighten the self-perception of each organization that was highlighted through social media. This further facilitated better participation and helped people to be more enthusiastic in the events that the city organized.

In 2014, five representatives or deputy representatives belonging to different civil society groups with active Facebook pages (*Childcare Rest Station & Relaxation Oasis*, *Clothing and Shelter Research Group*, *Tsukuba Children's Theater*, *Group for Enjoying and Playing with Kanji*, and *Tsukuba Assertiveness Society*) conducted an extended discussion during the second annual get-together event organized by the Tsukuba City Hall and this resulted in the establishment of a new group named *Tsukubano Ichinen wo Tanoshimu kai* (Let's enjoy Tsukuba year round). This was the first example in Tsukuba where the online and offline synergy resulted in producing a brand new endeavor, organized through the civil society organizations. However, insufficient funding for this new group along with the pre-existing groups becoming increasingly busy resulted in this new group to postpone new activities for the time being.

Two years later, in 2016, three more new groups were formed. The 2016 forum, saw the formation of a group named *UD Work*. Two therapists had decided to participate in the forum after seeing a notice about the forum on the Tsukuba Civic Activities Cyber-Square Facebook Page. However, neither of them were avid Facebook users. Their participation through social media was limited solely to the viewing of content. Through their encounter at the forum, they reaffirmed their shared interests and decided to begin working together. They formed an official group after realizing that to do so would be advantageous when they were applying for funding from Tsukuba City.

UD Work is comprised of two occupational therapists in their late thirties. With the objective of making it easier for patients who have finished their rehabilitation to proactively participate in outings, they visit and build new relationships at various places on a monthly basis. With no restrictions regarding age or type of disability, they aim for relaxed interactions which transcend both generational and medical boundaries. Specifically, they recruit participants for their outings from places such as the hospital and care center where they work and groups such as the *Chat Group for Families of Young Patients*. They now make use of their Facebook page to proactively share information and recruit participants.

The example of *UD Work* illustrates that sharing information through social media and raising awareness of relevant issues on the Tsukuba Civic Activities Cyber-Square Facebook Page and also through face-to-face encounters at the forum resulted in the formation of an entirely new group for the promotion of citizen activities. The interaction of offline and online spheres and information of actual citizen activities had the effect of expanding both the reach of activities and the loose networks involved.

The 2016 forum also harvested two other strong ties among civil society groups and individuals interested in civic activities. During the 2016 forum, two groups, the *Rescue Classroom for Kids* and the *Fun! Mysterious? Experiment Corps*, interacted with each other in the comments of the post. As both groups are quite active on the Tsukuba Civic Activities Cyber-Square Facebook Page, they were both already aware of each other, but the post about the forum provided the impetus for more direct online communication between these two groups.

As the forum sparked the start of loose interactions between the groups, the *Tsukuba Gakuin University Juggling Club*, which was invited to be an ice-breaker at the beginning of the forum also participated in the event and interacted with participants. One of the groups with which they interacted was the *Chat Group for Families of Young Patients*. Following the forum, the chat group shared an article about the Tsukuba Gakuin University Off-Campus Program on the Tsukuba Civic Activities Cyber-Square Facebook page. The representative of the chat group wrote in the comments for this post—"I had only heard about this from others so it was very useful to be able to read this article." This was the initial point for interaction between the *Tsukuba Gakuin University Juggling Club* and the chat group, with the *Tsukuba Gakuin University Juggling Club* later making a presentation of their activities for the chat group.

The interaction between the chat group and juggling club developed into further interactions between the chat group and the Tsukuba Gakuin University Off-Campus Program. The chat group made a request to Tsukuba Gakuin University and began to interact with students other than those involved with the *Tsukuba Gakuin University Juggling Club*. For instance, they participated in events such as the summer festival and Christmas party with students from the university. A participating student commented—"This summer festival was a collaborative production between the adults who planned it and the children who played there. Even in the very unique hospital environment, I was extremely pleased and greatly encouraged that the children had such a great time. If there is a chance to help out in this kind of event in the future, I would very much like to participate." The chat group posts these activities on Facebook and also shares other information and awareness about relevant issues. Citizens who have seen the posts have made comments such as "I'll definitely come next year … and help out as well!"

One common concern in Japan is how to encourage more young people to join and participate in civil society organizations that are already active. This may be a common feature in Japan and other nations, of course; however, new and constant participation is among senior citizens or older people and the organizations are in need of young people to join and participate. This example shows how this "barrier" can be conquered. We can observe how the loose interactions that began as a result of the gathering gradually expanded. Further, posting details of the interactions on social media generated new participants and new encounters.

These examples illustrate how sharing of information and promotion of awareness of relevant issues on the Tsukuba Civic Activities Cyber-Square Facebook Page combined with face-to-face encounters at the forum have resulted in a *regeneration* of civil society, or the formation of groups and the promotion of citizen activities. We observed how there is no necessity for individuals to be frequent users of social media for the procession from social media to group formation to be a successful one. The two individuals that formed *UD Work* after meeting face-to-face, for example, were participants of the Tsukuba Civic Activities Cyber-Square Facebook Page, but did not comment or share on the platform. Nevertheless, sharing information on social media and sharing awareness and perspectives on citizen activities can promote, in conjunction with face-to-face encounters, the formation of groups and loose networks.

The interaction of offline and online spheres, as seen in the independent transmission through Facebook of information about citizen activities, had the effect of expanding the reach of activities and the loose networks involved. We observed a series of flow-on effects as follows: (1) sharing of information and awareness of issues through social media; (2) face-to-face encounters; (3) formation of groups and loose networks; and (4) expansion of loose networks through social media.

These are the accomplishments of the Tsukuba Civic Activities Cyber-Square and the endeavors of social media usage at the local government level of Tsukuba City. We have shown how the efforts of the city using social media has actually transformed municipal workers and civil society and have led to the regeneration of new civic activities. We see a powerful regeneration of the Tsukuba City civil society being made through the use of social media.

References

Freberg, K. (2012). Intention to comply with crisis messages communicated via social media. *Public Relations Review, 38*(3), 416–421.

Park, J., Choi, H., & Park, S. M. (2011). Social media's impact on policy making. *SERI Quarterly*, 125–129.

Wendling, C., Radisch, J., & Jacobzone, S. (2013). *The use of social media in risk and crisis communication* (OECD working papers on Public Governance, No. 25). Paris: OECD Publishing. doi:10.1787/5k3v01fskp9s-en

Muneo Kaigo is an associate professor in the Faculty of Humanities and Social Sciences at the University of Tsukuba in Japan. He teaches courses in media communication in the Graduate School of Humanities and Social Sciences, media management in the Graduate School of Business Sciences and communication science in the College of Comparative Culture at the University of Tsukuba. He is currently leading a joint research project on social media uses among local municipalities in Japan with cooperation from the municipal government of Tsukuba and Intel Corporation Japan.

Sae Okura is an assistant professor in the Faculty of Humanities, Law and Economics at Mie University in Japan. She conducts research in politics and social policy studies. Her recent research interests focus on political participation by minority groups such as the disability community, sexual minorities, and single parent families. She analyzes quantitative survey data collected in Japan.

Appendix A: Number of Fans by Policy Areas (%)

M. Kaigo (ed.), *Social Media and Civil Society in Japan*,
DOI 10.1007/978-981-10-5095-4

	250	*251~500*	*501~1000*	*1001~2000*	*2001~3000*	*3001~4000*	*4001~5000*	*5001~6000*	*6001~7000*	*7001~8000*	*8001~9000*	*9001~*	*N*
Agriculture, forestry, and fisheries	23.8	14.3	33.3	19.0	9.5	0.0	0.0	0.0	0.0	0.0	0.0	0.0	21
Animal welfare	0.0	0.0	0.0	0.0	25.0	50.0	0.0	0.0	0.0	0.0	0.0	25.0	4
Childcare, education, and learning	20.0	16.0	36.0	20.0	0.0	4.0	0.0	0.0	4.0	0.0	0.0	0.0	25
Civic engagement	0.0	50.0	0.0	16.7	33.3	0.0	0.0	0.0	0.0	0.0	0.0	0.0	6
Community development	5.1	10.2	16.6	22.9	12.1	6.4	7.6	3.2	3.8	1.9	1.3	8.9	157
Consumer safety	100.0	0.0	0.0	0.0	0.0	0.0	0.0	0.0	0.0	0.0	0.0	0.0	2
Disaster prevention	20.0	10.0	20.0	20.0	20.0	0.0	0.0	10.0	0.0	0.0	0.0	0.0	10
Election	66.7	33.3	0.0	0.0	0.0	0.0	0.0	0.0	0.0	0.0	0.0	0.0	3
Employment	26.7	33.3	40.0	0.0	0.0	0.0	0.0	0.0	0.0	0.0	0.0	0.0	15
Enterprise	42.9	0.0	0.0	14.3	28.6	0.0	0.0	0.0	14.3	0.0	0.0	0.0	7
Entertainment	17.2	17.2	10.3	31.0	6.9	3.4	3.4	3.4	0.0	3.4	0.0	3.4	29
Environment	57.1	21.4	14.3	7.1	0.0	0.0	0.0	0.0	0.0	0.0	0.0	0.0	14
Foods	11.1	11.1	11.1	33.3	11.1	0.0	0.0	0.0	0.0	0.0	0.0	22.2	9
Gender equality	80.0	0.0	0.0	20.0	0.0	0.0	0.0	0.0	0.0	0.0	0.0	0.0	5
Health care and sports	36.4	27.3	18.2	9.1	0.0	0.0	0.0	9.1	0.0	0.0	0.0	0.0	11
Industrial development	0.0	42.9	42.9	14.3	0.0	0.0	0.0	0.0	0.0	0.0	0.0	0.0	7

International relations	50.0	50.0	0.0	0.0	0.0	0.0	0.0	0.0	0.0	0.0	0.0	0.0	2
Land, infrastructure, and transport	27.3	9.1	27.3	9.1	0.0	9.1	0.0	9.1	0.0	0.0	0.0	9.1	11
Multicultural coexistence	0.0	66.7	33.3	0.0	0.0	0.0	0.0	0.0	0.0	0.0	0.0	0.0	3
Public safety	33.3	0.0	0.0	0.0	33.3	33.3	0.0	0.0	0.0	0.0	0.0	0.0	3
Public Relations	8.3	0.0	12.5	16.7	16.7	8.3	0.0	8.3	4.2	0.0	8.3	16.7	24
Social welfare	50.0	0.0	50.0	0.0	0.0	0.0	0.0	0.0	0.0	0.0	0.0	0.0	4
Tourism	3.8	5.7	18.9	22.6	7.5	7.5	1.9	5.7	3.8	3.8	3.8	15.1	53
Total	15.1	12.9	18.8	19.3	9.4	5.2	3.3	3.3	2.6	1.4	1.4	7.3	425

Appendix B: Number of Engagement by Policy Areas (%)

M. Kaigo (ed.), *Social Media and Civil Society in Japan*,
DOI 10.1007/978-981-10-5095-4

	0	*1~3*	*4~10*	*11~50*	*51~100*	*101~200*	*201~500*	*501~1000*	*1000~*	*N*
Agriculture, forestry, and fisheries	28.6	4.8	19.0	33.3	0.0	9.5	4.8	0.0	0.0	21
Animal welfare	0.0	0.0	0.0	25.0	0.0	25.0	25.0	25.0	0.0	4
Childcare, education, and learning	12.0	12.0	8.0	28.0	20.0	16.0	0.0	4.0	0.0	25
Civic engagement	16.7	33.3	0.0	16.7	16.7	16.7	0.0	0.0	0.0	6
Community development	7.6	12.7	7.0	14.6	10.2	9.6	22.3	10.2	5.7	157
Consumer safety	50.0	50.0	0.0	0.0	0.0	0.0	0.0	0.0	0.0	2
Disaster prevention	30.0	10.0	0.0	20.0	20.0	0.0	10.0	10.0	0.0	10
Election	33.3	33.3	0.0	33.3	0.0	0.0	0.0	0.0	0.0	3
Employment	13.3	33.3	40.0	13.3	0.0	0.0	0.0	0.0	0.0	15
Enterprise	57.1	0.0	0.0	0.0	14.3	28.6	0.0	0.0	0.0	7
Entertainment	20.7	24.1	3.4	17.2	6.9	6.9	13.8	3.4	3.4	29
Environment	42.9	7.1	14.3	28.6	7.1	0.0	0.0	0.0	0.0	14
Foods	33.3	33.3	0.0	0.0	11.1	0.0	0.0	0.0	22.2	9
Gender equality	0.0	20.0	40.0	20.0	20.0	0.0	0.0	0.0	0.0	5
Health care and sports	36.4	27.3	0.0	27.3	0.0	0.0	0.0	9.1	0.0	11
Industrial development	14.3	14.3	14.3	57.1	0.0	0.0	0.0	0.0	0.0	7
International relations	50.0	0.0	0.0	50.0	0.0	0.0	0.0	0.0	0.0	2
Land, infrastructure, and transport	18.2	18.2	0.0	18.2	9.1	9.1	9.1	0.0	18.2	11
Multicultural coexistence	0.0	0.0	33.3	33.3	0.0	33.3	0.0	0.0	0.0	3
Public safety	33.3	0.0	0.0	0.0	0.0	0.0	33.3	33.3	0.0	3
Public relations	12.5	4.2	0.0	0.0	4.2	8.3	33.3	20.8	16.7	24
Social welfare	0.0	25.0	0.0	50.0	0.0	0.0	25.0	0.0	0.0	4
Tourism	17.0	11.3	3.8	15.1	9.4	11.3	5.7	11.3	15.1	53
Total	16.2	14.1	7.5	17.6	8.7	8.7	13.2	7.8	6.1	425

Appendix C: Overview of History of the "Tsukuba Civic Activities Cyber-Square" Facebook Page

1	Launch time of the Facebook page	1-Feb-12
2	Facebook group page manager	Division for volunteering support policy of the municipal government of Tsukuba, Ibaraki, Japan
3	Main target area	Tsukuba, Ibaraki, Japan
4	Objectives	To activate social network within a community and promote more civic engagement by developing information infrastructures
5	Accumulated "Likes" of the Facebook page (as of 17 April 2016)	2604
6	Gender of users by age (as of 17 April 2016)	–24: women 3%, men 4% 25–34: women 11%, men 14% 35–44: women 12%, men 20% 45–54: women 8%, men 13% 55–64: women 3%, men 6% 65+: women 1%, men 4%
7	Residential Profiles of users (as of 27 January 2014)	Tsukuba, Ibaraki, Japan: 54.2% Chiyoda-ku, Tokyo, Japan: 8.7% Tsuchiura, Ibaraki, Japan: 4.4% Mito, Ibaraki, Japan 2.9% etc.

M. Kaigo (ed.), *Social Media and Civil Society in Japan*,
DOI 10.1007/978-981-10-5095-4

Appendix D: TMin., Max., Avg. and Median Betweenness Centrality and Closeness Centrality

	NHA (12 June 2015)	*Tornado in Hojo, Ibaraki (12 June 2012)*	*Social welfare (25 January 2016)*	*Volunteering (28 January 2016)*
Minimum betweenness centrality	0	0	0	0
Maximum betweenness centrality	63	220.176	42	0
Average betweenness centrality	10.356	20.217	1.826	0
Median betweenness centrality	0	0	0	0
Minimum closeness centrality	0.008	0.011	0.023	0.024
Maximum closeness centrality	0.014	0.022	0.045	0.024
Average closeness centrality	0.012	0.016	0.043	0.024
Median closeness centrality	0.013	0.016	0.043	0.024

M. Kaigo (ed.), *Social Media and Civil Society in Japan*,
DOI 10.1007/978-981-10-5095-4

Index

M. Kaigo (ed.), *Social Media and Civil Society in Japan*,
DOI 10.1007/978-981-10-5095-4

X

Y

Z

Kolibri 360 GmbH
Ferdinand-Jühlke-Straße 7
99095 Erfurt, Deutschland
produktsicherheit@kolibri360.de